Lecture Notes in Physics

24

R. F. Snipes
Bowling Green State University, Bowling Green, Ohio/USA

Statistical Mechanical Theory of the Electrolytic Transport of Non-electrolytes

Springer-Verlag Berlin Heidelberg GmbH 1973

ISBN 978-3-540-06566-1 ISBN 978-3-540-37862-4 (eBook)
DOI 10.1007/978-3-540-37862-4

Originally published by Springer-Verlag Berlin Heidelberg New York in 1973.

Table of Contents

Abstract

When an electric current is passed through an electrolytic solution to which a non-electrolyte has been added, the non-electrolyte usually undergoes transport. In this paper a statistical mechanical theory of this electrolytic transport of non-electrolytes is given. The formal theory of transport processes developed by Kirkwood, Irving, Bearman, and Dahler is extended to multi-component fluids composed of rigid impenetrable molecules having rotational as well as translational degrees of freedom. Then, using a continuum model, the steady state equations of motion and continuity used by Debye, Hückel, Falkenhagen, Onsager and Fuoss in their theories of irreversible processes in dilute electrolytic solutions are extended so as to be applicable to dilute solutions with spherical solute particles having rotational as well as translational degrees of freedom. Finally, a simple continuum model is used to represent the solution containing non-electrolyte molecules and electrolyte ions placed in a constant homogeneous electric field; and with an approximate theory valid at infinite dilution, the mass transport ratio for an ideal dipolar non-electrolyte is calculated. Comparisons with experiments are made.

I. Introduction [1,2]

When an electric current is passed through an electrolytic solution to which a non-electrolyte has been added, the non-electrolyte usually undergoes transport. Early experimenters[3] assumed the added non-electrolyte to be uninfluenced by the passage of the electric current. However, Longsworth,[4] using the moving boundary method, showed conclusively that the added non-electrolyte is not motionless. Indeed, Longsworth found that for a given electrolyte and given concentrations, different non-electrolytes move with different velocities. The purpose of this paper is to present a statistical mechanical theory of this electrolytic transport of non-electrolytes.

The simplest "realistic" model for a non-electrolyte molecule is a rigid body having an arbitrary charge and mass distribution. Consequently, in Section II of this paper, the formal theory of transport processes[5-9] developed by Kirkwood, Irving, Bearman, and Dahler is extended to multicomponent fluids composed of rigid impenetrable molecules (or ions) having rotational as well as translational degrees of freedom. A number of relevant macroscopic quantities are represented as averages over molecular distribution functions; non-equilibrium distribution functions are expressed in terms of first order perturbations to the equilibrium (canonical ensemble) distribution functions; and the hydrodynamic equations of continuity and motion are derived.

After establishing these formal results, the second objective of the statistical mechanical theory of transport is the explicit determination of the non-equilibrium perturbations in the distribution functions. A rigorous treatment of this problem is not possible, at present. However, when details of solvent structure are neglected, i.e., when the solvent is represented as a viscous, incompressible, dielectric continuum; approximate calculations can be carried out for dilute solutions of rigid spherical particles in a constant external field and at a constant uniform temperature. In this paper, such calculations are based on the steady state equations of motion and continuity used by Debye, Hückel, Falkenhagen, Onsager, and Fuoss[10] in their theories of irreversible processes in dilute electrolytic solutions; and on extensions of these equations to dilute solutions with spherical solute particles having rotational as well as translational degrees of freedom. This theory is presented in Section III.

Having developed a sufficiently general statistical mechanical theory of transport phenomena, we now specialize to the problem at hand: the electrolytic transport of non-electrolytes. This theory is given in Section IV.

The dilute solution containing non-electrolyte molecules and electrolyte ions is represented by a model[11] whereby: (1) the solvent is a viscous, incompressible, dielectric continuum; (2) the non-electrolyte molecules, present at extremely low concentrations so that their mutual interactions can be neglected, are rigid spheres having low dielectric constants; arbitrary, but discrete, charge distributions; and the center of mass of each sphere at its center; (3) the electrolyte ions are rigid spheres having spherically symmetrical charge and mass distributions; and dielectric constants the same as that of the solvent; and (4) in addition to the long-range electrostatic forces, short-range forces, possibly just ion-cavity repulsive forces, are operative[12] between the non-electrolyte molecules and the electrolyte ions.

The motion of neutral non-electrolyte molecules attending the passage of the electric current through the electrolytic solution (viz., when the solution is placed in a constant homogeneous electric field and maintained at a constant uniform temperature) can be described either in terms of the non-electrolyte mobility or in terms of the non-electrolyte mass transport ratio.[13]

Using the afore-mentioned model with the charge distribution of the non-electrolyte molecule represented as a centrally located ideal dipole, theoretical expressions valid at infinite dilution are obtained for these two quantities. These calculations are similar to those of the Onsager-Fuoss theory[14] of the conductance of electrolytes. Following their procedure, three origins of translatory force acting on a reference non-electrolyte molecule are considered. The first, called the relaxation force, is due to the asymmetry which the external electric field produces in the distribution of electrolyte ions about the non-electrolyte molecule. Because of this asymmetry, the forces between the ions and the non-electrolyte molecule are no longer in balance; and there is an average resultant translatory force on the molecule. The second, called the kinetic force, is also due to the asymmetry in the distribution of electrolyte ions about the non-electrolyte molecule. This effect arises through an imbalance of collisions between the ions and the non-electrolyte molecule; or, alternately, through an imbalance of the osmotic pressure on the reference non-electrolyte molecule. In order to obtain explicit expressions for these forces the perturbation in the (non-electrolyte molecule, electrolyte ion)-pair correlation function is needed. This function is obtained by solving the equation of continuity in the (non-electrolyte molecule, electrolyte ion)-pair configuration space. The third and final force gives rise to the electrophoretic velocity of the non-electrolyte molecule. This velocity results from fluid motion in the neighborhood of the non-electrolyte molecule. In determining this velocity, we assume that our model obeys Stokes' law.

The calculations carried out in Section IV are valid only at infinite dilution. In the limiting "law" for the non-electrolyte mobility; terms linear in the electrolyte concentration, but independent of the non-electrolyte concentration, appear. The limiting "law" for the mass transport ratio contains terms linear in the non-electrolyte concentration but independent of the electrolyte concentration.[15]

Finally, in Section V, the mass transport ratios predicted by the theory for infinitely dilute solutions are compared with experimental values.

II. Statistical Mechanical Theory of Transport Processes for Multi-component Poly-atomic Fluids

We shall consider a classical fluid system containing $\underline{N}$ molecules (or ions), which are not chemically reacting, in a volume $\underline{V}$. There are $\underline{N}_\alpha$ molecules of species α where α runs from 1 to σ. The individual molecules of each species are labeled separately, so that the molecules of species α are numbered from 1 to $\underline{N}_\alpha$.

Generalized Coordinates and Momenta

Each molecule has six degrees of freedom: three translational degrees of freedom and three rotational degrees of freedom. Thus we may specify the configuration of molecule $\underline{i}$ of species α by the vector $\underset{\sim}{q}_{\alpha\underline{i}} = (\underline{q}_{\alpha\underline{i}1}, \underline{q}_{\alpha\underline{i}2}, \ldots, \underline{q}_{\alpha\underline{i}6})$ chosen such that the vector $\underset{\sim}{r}_{\alpha\underline{i}} = (\underline{q}_{\alpha\underline{i}1}, \underline{q}_{\alpha\underline{i}2}, \underline{q}_{\alpha\underline{i}3})$ specifies the position of its center of mass and the vector $\underset{\sim}{R}_{\alpha\underline{i}} = (\underline{q}_{\alpha\underline{i}4}, \underline{q}_{\alpha\underline{i}5}, \underline{q}_{\alpha\underline{i}6})$ specifies its orientation. We specify the velocity of molecule $\underline{i}$ of species α by the vector $\dot{\underset{\sim}{q}}_{\alpha\underline{i}} = (\dot{\underline{q}}_{\alpha\underline{i}1}, \dot{\underline{q}}_{\alpha\underline{i}2}, \ldots, \dot{\underline{q}}_{\alpha\underline{i}6})$ such that $\dot{\underset{\sim}{r}}_{\alpha\underline{i}}$ specifies its linear velocity and $\dot{\underset{\sim}{R}}_{\alpha\underline{i}}$ specifies its angular velocity. Upon introducing generalized momenta $\underline{p}_{\alpha\underline{i}\underline{j}}$ conjugate to the generalized coordinates $\underline{q}_{\alpha\underline{i}\underline{j}}$, we may specify the momenta of molecule $\underline{i}$ of species α by the vector $\underset{\sim}{p}_{\alpha\underline{i}} = (p_{\alpha\underline{i}1}, p_{\alpha\underline{i}2}, \ldots, p_{\alpha\underline{i}6})$ such that the vector $\underset{\sim}{m}_{\alpha\underline{i}} = (p_{\alpha\underline{i}1}, p_{\alpha\underline{i}2}, p_{\alpha\underline{i}3})$ specifies its linear momentum and the vector $\underset{\sim}{M}_{\alpha\underline{i}} = (p_{\alpha\underline{i}4}, p_{\alpha\underline{i}5}, p_{\alpha\underline{i}6})$ specifies its rotational momenta. We also use the notation $\underset{\sim}{q}_{\alpha\underline{i}} = (\underset{\sim}{r}_{\alpha\underline{i}}, \underset{\sim}{R}_{\alpha\underline{i}})$ for the complete set of configuration coordinates of molecule $\underline{i}$ of species α ; the notation $\dot{\underset{\sim}{q}}_{\alpha\underline{i}} = (\dot{\underset{\sim}{r}}_{\alpha\underline{i}}, \dot{\underset{\sim}{R}}_{\alpha\underline{i}})$ for the complete set of generalized velocities; the notation $\underset{\sim}{p}_{\alpha\underline{i}} = (\underset{\sim}{m}_{\alpha\underline{i}}, \underset{\sim}{M}_{\alpha\underline{i}})$ for the complete set of momenta coordinates; and $(\underset{\sim}{q}_{\alpha\underline{i}}, \underset{\sim}{p}_{\alpha\underline{i}})$ for a point in the phase space of molecule $\underline{i}$ of species α. When referring to an arbitrary molecule of species α, we drop the subscript $\underline{i}$.

For the generalized coordinates which serve to locate the position of the center of mass of an arbitrary molecule of species α relative to a point fixed in space and chosen as the origin, we may choose either rectangular Cartesian coordinates so that $\underset{\sim}{r}_{\alpha} = (x_{\alpha}, y_{\alpha}, z_{\alpha})$ and $\underset{\sim}{m}_{\alpha} = (p_{x_{\alpha}}, p_{y_{\alpha}}, p_{z_{\alpha}})$; or spherical polar coordinates[16] such that $\underset{\sim}{r}_{\alpha} = (r_{\alpha}, \theta_{\alpha}, \varphi_{\alpha})$ and $\underset{\sim}{m}_{\alpha} = (p_{r_{\alpha}}, p_{\theta_{\alpha}}, p_{\varphi_{\alpha}})$. If we associate with the rectangular Cartesian coordinates the constant space-fixed orthonormal basic vectors $\underset{\sim}{i} = (1, 0, 0)$, $\underset{\sim}{j} = (0, 1, 0)$ and $\underset{\sim}{k} = (0, 0, 1)$, we may write $\underset{\sim}{r}_{\alpha} = x_{\alpha}\,\underset{\sim}{i} + y_{\alpha}\,\underset{\sim}{j} + z_{\alpha}\,\underset{\sim}{k}$. Similarly, if we introduce the orthonormal basic vectors $\underset{\sim}{1}_{r_{\alpha}}$,

$\underset{\sim}{1}_{\theta_\alpha}$, and $\underset{\sim}{1}_{\varphi_\alpha}$ associated with the spherical polar coordinates $(\underline{r}_\alpha, \theta_\alpha, \varphi_\alpha)$ and defined by the equations

$$\begin{aligned} \underset{\sim}{1}_{\underline{r}_\alpha} &= \sin\theta_\alpha \cos\varphi_\alpha \underset{\sim}{i} + \sin\theta_\alpha \sin\varphi_\alpha \underset{\sim}{j} + \cos\theta_\alpha \underset{\sim}{k} \\ \underset{\sim}{1}_{\theta_\alpha} &= \cos\theta_\alpha \cos\varphi_\alpha \underset{\sim}{i} + \cos\theta_\alpha \sin\varphi_\alpha \underset{\sim}{j} - \sin\theta_\alpha \underset{\sim}{k} \\ \underset{\sim}{1}_{\varphi_\alpha} &= -\sin\varphi_\alpha \underset{\sim}{i} + \cos\varphi_\alpha \underset{\sim}{j} , \end{aligned} \tag{1}$$

we may write $\underset{\sim}{r}_\alpha = \underline{r}_\alpha \underset{\sim}{1}_{\underline{r}_\alpha}$.

For the generalized coordinates which serve to specify the orientation of an arbitrary molecule of species α, we choose the Eulerian angles $\underset{\sim}{R}_\alpha = (\Phi_\alpha, \Theta_\alpha, \Xi_\alpha)$ whereby one can transform the set of constant orthonormal basic vectors $\{\underset{\sim}{i}, \underset{\sim}{j}, \underset{\sim}{k}\}$, bound to the center of mass of the molecule and fixed in space, into the set of orthonormal basic vectors $\{\underset{\sim}{i}_\alpha, \underset{\sim}{j}_\alpha, \underset{\sim}{k}_\alpha\}$, not only bound to the center of mass of the molecule of species α but also affixed to the molecule so as to rotate with it, by the successive rotations:

(1) a rotation by an angle Φ_α, where $0 \le \Phi_\alpha < 2\pi$, counterclockwise about $\underset{\sim}{k}$;

(2) a rotation by an angle Θ_α, where $0 \le \Theta_\alpha \le \pi$, counterclockwise about $\underset{\sim}{1}_{\Phi_\alpha} = -\sin\Phi_\alpha \underset{\sim}{i} + \cos\Phi_\alpha \underset{\sim}{j}$, bound to the center of mass of the molecule; and

(3) a rotation by an angle Ξ_α, where $0 \le \Xi_\alpha < 2\pi$,

counterclockwise about $\underline{1}_{R_\alpha} = \sin\Theta_\alpha \cos\Phi_\alpha\, \underline{i} + \sin\Theta_\alpha \sin\Phi_\alpha\, \underline{j} + \cos\Theta_\alpha\, \underline{k}$ also bound to the center of mass of the molecule.

We thus write $\underline{M}_\alpha = (p_{\Phi_\alpha}, p_{\Theta_\alpha}, p_{\Xi_\alpha})$. The orthonormal basic vectors $\underline{i}_\alpha, \underline{j}_\alpha, \underline{k}_\alpha$ can be expanded in terms of the constant orthonormal basic vectors $\underline{i}$, $\underline{j}$, and $\underline{k}$, as follows:

$$\begin{aligned}
\underline{i}_\alpha &= (\cos\Xi_\alpha \cos\Theta_\alpha \cos\Phi_\alpha - \sin\Xi_\alpha \sin\Phi_\alpha)\, \underline{i} \\
&\quad + (\cos\Xi_\alpha \cos\Theta_\alpha \sin\Phi_\alpha + \sin\Xi_\alpha \cos\Phi_\alpha)\, \underline{j} \\
&\quad - (\cos\Xi_\alpha \sin\Theta_\alpha)\, \underline{k} \\
\underline{j}_\alpha &= (-\sin\Xi_\alpha \cos\Theta_\alpha \cos\Phi_\alpha - \cos\Xi_\alpha \sin\Phi_\alpha)\, \underline{i} \\
&\quad + (-\sin\Xi_\alpha \cos\Theta_\alpha \sin\Phi_\alpha + \cos\Xi_\alpha \cos\Phi_\alpha)\, \underline{j} \\
&\quad + (\sin\Xi_\alpha \sin\Theta_\alpha)\, \underline{k} \\
\underline{k}_\alpha &= \sin\Theta_\alpha \cos\Phi_\alpha\, \underline{i} + \sin\Theta_\alpha \sin\Phi_\alpha\, \underline{j} \\
&\quad + \cos\Theta_\alpha\, \underline{k}\,.
\end{aligned} \tag{2}$$

This definition of the Eulerian angles is not the usual one.[17] However, it has the advantage that the two Eulerian angles Φ_α and Θ_α, together with R_α, the distance from the center of mass of the molecule to some point $R_\alpha\, \underline{k}_\alpha$, become the spherical polar coordinates $(R_\alpha, \Theta_\alpha, \Phi_\alpha)$ of that point with respect to the constant orthonormal basic vectors $\underline{i}$, $\underline{j}$, and $\underline{k}$ bound to the center of mass of the molecule. Moreover, the orthonormal basic vectors $\underline{1}_{R_\alpha}$, $\underline{1}_{\Theta_\alpha}$, and $\underline{1}_{\Phi_\alpha}$ associated with these spherical polar coordinates can be obtained by setting Ξ_α equal to zero in equations (2) yielding as the result:

$$\begin{aligned} \mathbf{1}_{R_\alpha} &= \mathbf{k}_\alpha = \sin\Theta_\alpha \cos\Phi_\alpha\, \mathbf{i} + \sin\Theta_\alpha \sin\Phi_\alpha\, \mathbf{j} \\ &\quad + \cos\Theta_\alpha\, \mathbf{k} \\ \mathbf{1}_{\Theta_\alpha} &= \cos\Theta_\alpha \cos\Phi_\alpha\, \mathbf{i} + \cos\Theta_\alpha \sin\Phi_\alpha\, \mathbf{j} \\ &\quad - \sin\Theta_\alpha\, \mathbf{k} \\ \mathbf{1}_{\Phi_\alpha} &= -\sin\Phi_\alpha\, \mathbf{i} + \cos\Phi_\alpha\, \mathbf{j}\,. \end{aligned} \tag{3}$$

If the molecule of species α has an <u>axis of symmetry</u>; and if the unit vector $\mathbf{k}_\alpha$ is chosen so as to lie along this axis; then the physically distinguishable orientations of the molecule are uniquely determined by $\mathbf{k}_\alpha$ or by the two Eulerian angles Θ_α and Φ_α. In particular, if the molecule has the dipole moment $\boldsymbol{\mu}$, we might choose $\mathbf{k}_\alpha$ so that $\boldsymbol{\mu} = \mu\, \mathbf{k}_\alpha$. Then the vector $\boldsymbol{\mu}$ would specify the orientation of the molecule and give, in addition, the magnitude μ of its electric dipole moment.

The <u>kinetic energy</u> T_α of a molecule of species α divides naturally into contributions from the translation of its center of mass and from the rotation about its center of mass, i.e.,

$$T_\alpha = \frac{1}{2} m_\alpha\, \dot{\mathbf{r}}_\alpha \cdot \dot{\mathbf{r}}_\alpha + \frac{1}{2} \boldsymbol{\omega}_\alpha \cdot \mathbf{L}_\alpha \tag{4}$$

where m_α is the mass of the molecule of species α; where $\dot{\mathbf{r}}_\alpha = \dot{x}_\alpha \mathbf{i} + \dot{y}_\alpha \mathbf{j} + \dot{z}_\alpha \mathbf{k} = \dot{r}_\alpha \mathbf{1}_{r_\alpha} + r_\alpha \dot{\theta}_\alpha \mathbf{1}_{\theta_\alpha}$

$+ \underline{r}_\alpha \dot{\varphi}_\alpha \sin\theta_\alpha \, \underset{\sim}{1}_{\varphi_\alpha}$ is the translational velocity of the center of mass of the molecule of species α ; where $\underset{\sim}{\omega}_\alpha$ = $\dot{\underset{\sim}{R}}_\alpha = \dot{\Phi}_\alpha \underset{\sim}{k} + \dot{\Theta}_\alpha \underset{\sim}{1}_{\Phi_\alpha} + \dot{\Xi}_\alpha \underset{\sim}{k}_\alpha$ is the angular velocity of the molecule of species α ; and where $\underset{\sim}{L}_\alpha = \underset{\sim}{I}_\alpha \cdot \underset{\sim}{\omega}_\alpha$ is the angular momentum of the molecule of species α arising solely from its rotational motion about its center of mass. Here $\underset{\sim}{I}_\alpha$ is the moment of inertia tensor of the molecule of species α. The linear momentum of the molecule of species α is given by the vector $\underset{\sim}{m}_\alpha = \underline{m}_\alpha \dot{\underset{\sim}{r}}_\alpha = \underline{p}_{x_\alpha} \underset{\sim}{i} + \underline{p}_{y_\alpha} \underset{\sim}{j} + \underline{p}_{z_\alpha} \underset{\sim}{k}$ = $\underline{p}_{r_\alpha} \underset{\sim}{1}_{r_\alpha} + \frac{1}{\underline{r}_\alpha} \underline{p}_{\theta_\alpha} \underset{\sim}{1}_{\theta_\alpha} + \frac{1}{\underline{r}_\alpha \sin\theta_\alpha} \underline{p}_{\varphi_\alpha} \underset{\sim}{1}_{\varphi_\alpha}$ where $\underline{p}_{x_\alpha} = \underline{m}_\alpha \dot{\underline{x}}_\alpha$, $\underline{p}_{y_\alpha} = \underline{m}_\alpha \dot{\underline{y}}_\alpha$, $\underline{p}_{z_\alpha} = \underline{m}_\alpha \dot{\underline{z}}_\alpha$; and $\underline{p}_{r_\alpha}$ = $\underline{m}_\alpha \dot{\underline{r}}_\alpha$, $\underline{p}_{\theta_\alpha} = \underline{m}_\alpha \underline{r}_\alpha^2 \dot{\theta}_\alpha$, and $\underline{p}_{\varphi_\alpha} = \underline{m}_\alpha \underline{r}_\alpha^2 \dot{\varphi}_\alpha \sin^2\theta_\alpha$. Since the momentum $\underline{p}_j$ conjugate to a rotation coordinate $\underline{q}_j$ is the component of the angular momentum along the axis of rotation, the rotational momenta conjugate to the Eulerian angles $\underset{\sim}{R}_\alpha = (\Phi_\alpha, \Theta_\alpha, \Xi_\alpha)$ are given by $\underset{\sim}{M}_\alpha = (\underline{p}_{\Phi_\alpha}, \underline{p}_{\Theta_\alpha}, \underline{p}_{\Xi_\alpha})$ where $\underline{p}_{\Phi_\alpha} = \underset{\sim}{L}_\alpha \cdot \underset{\sim}{k}$, $\underline{p}_{\Theta_\alpha} = \underset{\sim}{L}_\alpha \cdot \underset{\sim}{1}_{\Phi_\alpha}$ and $\underline{p}_{\Xi_\alpha} = \underset{\sim}{L}_\alpha \cdot \underset{\sim}{k}_\alpha$.

The <u>element of extension</u>[18] in the configuration space of a molecule of species α is $\underline{m}_\alpha^{3/2} \underline{I}_\alpha^{3/2} \underline{d}^6 \underset{\sim}{q}_\alpha$ where

where I_α^3 is a constant depending only on the mass distribution within the molecule; and where $d^6 q_\alpha = d^3 r_\alpha \, d^3 R_\alpha$, with $d^3 r_\alpha = dx_\alpha \, dy_\alpha \, dz_\alpha = r_\alpha^2 \sin \theta_\alpha \, dr_\alpha \, d\theta_\alpha \, d\varphi_\alpha$ being the volume element in the position space of the molecule of species α and with $d^3 R_\alpha = \sin \Theta_\alpha \, d\Phi_\alpha \, d\Theta_\alpha \, d\Xi_\alpha$ being the volume element in the orientation space of the molecule of species α. Similarly, the element of extension in the momentum space of a molecule of species α is $m_\alpha^{-3/2} \, I_\alpha^{-3/2} \, d^6 p_\alpha$ where $d^6 p_\alpha = d^3 m_\alpha \, d^3 M_\alpha$, with $d^3 m_\alpha = dp_{x_\alpha} \, dp_{y_\alpha} \, dp_{z_\alpha}$ $= (r_\alpha^2 \sin \theta_\alpha)^{-1} \, dp_{r_\alpha} \, dp_{\theta_\alpha} \, dp_{\varphi_\alpha}$ being the volume element in the linear momentum space of the molecule of species α and with $d^3 M_\alpha = (\sin \Theta_\alpha)^{-1} \, dp_{\Phi_\alpha} \, dp_{\Theta_\alpha} \, dp_{\Xi_\alpha}$ being the volume element in the rotational momenta space of the molecule of species α. Finally, the element of extension in the phase space of the molecule of species α is $d^6 q_\alpha \, d^6 p_\alpha =$ $d^3 r_\alpha \, d^3 R_\alpha \, d^3 m_\alpha \, d^3 M_\alpha$.

The configuration of the system of N molecules can be specified by giving the generalized coordinates $q =$ $(q_{11}, q_{12}, \ldots, q_{\sigma N_\sigma})$ where the vector $q_{\alpha i} = (q_{\alpha i 1}, \ldots, q_{\alpha i 6})$ specifies the configuration of molecule i of species α. Similarly, we introduce the generalized velocities $\dot{q} =$ $(\dot{q}_{11}, \dot{q}_{12}, \ldots, \dot{q}_{\sigma N_\sigma})$ as well as the generalized momenta $p = (p_{11}, p_{12}, \ldots, p_{\sigma N_\sigma})$. Thus we may use the notation

(q, p) for a point in the phase space of the entire system of N molecules. The kinetic energy of the system is given by

$$T = \sum_{\alpha=1}^{\sigma} \sum_{i=1}^{N_\alpha} T_{\alpha i} \tag{5}$$

where $T_{\alpha i}$ is the kinetic energy of molecule i of species α as given by equation (4) provided the subscript α is replaced by the subscript αi. The element of extension in the configuration space of this system of N molecules is $\prod_{\alpha=1}^{\sigma} (m_\alpha I_\alpha)^{3/2\, N_\alpha} \; dq$ where $dq = d^6q_{11}\; d^6q_{12} \cdots d^6q_{\sigma N_\sigma}$ with $d^6q_{\alpha i} = d^3r_{\alpha i}\; d^3R_{\alpha i}$. Correspondingly, the element of extension in momentum space is $\prod_{\alpha=1}^{\sigma} (m_\alpha I_\alpha)^{-3/2\, N_\alpha} \; dp$ where $dp = d^6p_{11}\; d^6p_{12} \cdots d^6p_{\sigma N_\sigma}$ with $d^6p_{\alpha i} = d^3m_{\alpha i}\; d^3M_{\alpha i}$. Finally, the element of extension in the phase space of the entire system is $dq\, dp = \prod_{\alpha=1}^{\sigma} \prod_{i=1}^{N_\alpha} d^6q_{\alpha i}\; d^6p_{\alpha i}$.

Forces and Torques

We now assume that the **potential energy** U of the system (We will write $U(q)$ instead of U if we wish to emphasize that the potential energy of the system depends only on its configuration $\underset{\sim}{q}$.) is of the form

$$\underline{U}(\underset{\sim}{q}) = \sum_{\alpha=1}^{\sigma} \sum_{\underline{i}=1}^{N_\alpha} \underline{V}_\alpha^{(e)}(\underset{\sim}{q}_{\alpha\underline{i}}) + (1/2) \sum_{\alpha=1}^{\sigma} \sum_{\beta=1}^{\sigma} \sum_{\underline{i}=1}^{N_\alpha} \sum_{\underline{j}=1}^{N_\beta}{}_{\alpha\underline{i} \neq \beta\underline{j}} \underline{V}_{\alpha\beta}(\underset{\sim}{q}_{\alpha\underline{i}}, \underset{\sim}{q}_{\beta\underline{j}}) \tag{6}$$

where $\underline{V}_\alpha^{(e)}(\underset{\sim}{q}_{\alpha\underline{i}})$ is the potential energy of a molecule of species α in configuration $\underset{\sim}{q}_{\alpha\underline{i}}$ arising from conservative external fields; and where $\underline{V}_{\alpha\beta}(\underset{\sim}{q}_{\alpha\underline{i}}, \underset{\sim}{q}_{\beta\underline{j}})$ is the mutual potential energy of a pair of molecules, one of species α in configuration $\underset{\sim}{q}_{\alpha\underline{i}}$ and one of species β in configuration $\underset{\sim}{q}_{\beta\underline{j}}$. This pair potential energy is a function not only of $\underset{\sim}{r}_{\alpha\underline{i}\beta\underline{j}} = \underset{\sim}{r}_{\alpha\underline{i}} - \underset{\sim}{r}_{\beta\underline{j}}$, the vector connecting the centers of mass of the two molecules, but also of the orientations $\underset{\sim}{R}_{\alpha\underline{i}}$ and $\underset{\sim}{R}_{\beta\underline{j}}$ of the two molecules.

The generalized force $\underline{F}_{\alpha\underline{ik}}$ associated with $\underline{q}_{\alpha\underline{ik}}$, the $\underline{k}$th generalized coordinate used in specifying the configuration of the $\underline{i}$th molecule of species α, is given (for a conservative system) by

$$\underline{F}_{\alpha\underline{ik}} = - \frac{\partial \underline{U}}{\partial \underline{q}_{\alpha\underline{ik}}} .$$

If $\underline{d}\,\underline{q}_{\alpha\underline{ik}}$ represents a translation of the $\underline{i}$th molecule of species α in some given direction, then $\underline{F}_{\alpha\underline{ik}}$ represents the component of the total force on that molecule along the direction of translation. Consequently, the total force acting at the center of mass of the ith molecule of species α (in configuration $\underline{q}_{\alpha i} = (\underline{r}_{\alpha i}, \underline{R}_{\alpha i})$) for a fixed configuration $\underline{q}$ of all the molecules in the system is

$$\underline{F}_{\alpha i} = - \nabla_{\underline{r}_{\alpha i}} U \tag{7}$$

where

$$\nabla_{\underline{r}_{\alpha i}} = \underline{i} \frac{\partial}{\partial x_{\alpha i}} + \underline{j} \frac{\partial}{\partial y_{\alpha i}} + \underline{k} \frac{\partial}{\partial z_{\alpha i}} \tag{8}$$

is the gradient operator associated with the rectangular Cartesian coordinates $\underline{r}_{\alpha i} = (x_{\alpha i}, y_{\alpha i}, z_{\alpha i})$ of the center of mass of the ith molecule of species α, or alternately,

$$\begin{aligned} \nabla_{\underline{r}_{\alpha i}} = \underline{1}_{r_{\alpha i}} \frac{\partial}{\partial r_{\alpha i}} + \underline{1}_{\theta_{\alpha i}} \frac{1}{r_{\alpha i}} \frac{\partial}{\partial \theta_{\alpha i}} \\ + \underline{1}_{\varphi_{\alpha i}} \frac{1}{r_{\alpha i} \sin \theta_{\alpha i}} \frac{\partial}{\partial \varphi_{\alpha i}} \end{aligned} \tag{9}$$

is the gradient operator associated with its spherical polar coordinates $\underline{r}_{\alpha i} = (r_{\alpha i}, \theta_{\alpha i}, \varphi_{\alpha i})$. With the potential energy of the system given by equation (6), the total force $\underline{F}_{\alpha i}$ on the ith molecule of species α becomes

$$\underline{F}_{\alpha i}(\underline{q}) = \underline{F}_{\alpha i}^{(e)}(\underline{q}_{\alpha i}) + \sum_{\beta=1}^{\sigma} \sum_{\substack{j=1 \\ \beta j \neq \alpha i}}^{N_\beta} \underline{F}_{\beta j \alpha i}(\underline{q}_{\beta j}, \underline{q}_{\alpha i}) \tag{10}$$

where $\underline{F}_{\alpha i}^{(e)}(\underline{q}_{\alpha i}) = - \nabla_{\underline{r}_{\alpha i}} V_{\alpha}^{(e)}(\underline{q}_{\alpha i})$ is the force exerted on molecule $\underline{i}$ of species α by the external field; and where $\underset{\approx}{F}_{\beta j \alpha \underline{i}}(\underset{\sim}{q}_{\beta j}, \underset{\sim}{q}_{\alpha \underline{i}}) = -\nabla_{\underset{\sim}{r}_{\alpha \underline{i}}} V_{\alpha\beta}(\underset{\sim}{q}_{\alpha \underline{i}}, \underset{\sim}{q}_{\beta \underline{j}})$ is the force exerted on molecule $\underline{i}$ of species α by molecule $\underline{j}$ of species β. If together with the vector $\underset{\sim}{r}_{\alpha \underline{i} \beta \underline{j}} = \underset{\sim}{r}_{\alpha \underline{i}} - \underset{\sim}{r}_{\beta \underline{j}}$ we introduce relative coordinates (rectangular Cartesian or spherical polar) in position space having the center of mass $\underset{\sim}{r}_{\beta \underline{j}}$ of molecule $\underline{j}$ of species β as the origin; and if we introduce the gradient operator $\nabla_{\underset{\sim}{r}_{\alpha \underline{i} \beta \underline{j}}} = \nabla_{\underset{\sim}{r}_{\alpha \underline{i}}} = -\nabla_{\underset{\sim}{r}_{\beta \underline{j}}}$; then $\underset{\sim}{F}_{\beta \underline{j} \alpha \underline{i}} = -\underset{\sim}{F}_{\alpha \underline{i} \beta \underline{j}} = -\nabla_{\underset{\sim}{r}_{\alpha \underline{i} \beta \underline{j}}} V_{\alpha\beta}$.

The total <u>torque</u> (about the center of mass of the $\underline{i}^{th}$ molecule of species α as origin) acting on the $\underline{i}^{th}$ molecule of species α (in configuration $\underset{\sim}{q}_{\alpha \underline{i}} = (\underset{\sim}{r}_{\alpha \underline{i}}, \underset{\sim}{R}_{\alpha \underline{i}})$) for a fixed configuration $\underset{\approx}{q}$ of all the molecules in the system is

$$\underset{\sim}{T}_{\alpha \underline{i}} = -\nabla_{\underset{\sim}{R}_{\alpha \underline{i}}} \underline{U} \tag{11}$$

where

$$\nabla_{\underset{\sim}{R}_{\alpha \underline{i}}} = \underset{\sim}{1}_{R_{\alpha \underline{i}}} \frac{\partial}{\partial \Xi_{\alpha \underline{i}}} + \underset{\sim}{1}_{\Phi_{\alpha \underline{i}}} \frac{\partial}{\partial \Theta_{\alpha \underline{i}}} + \underset{\sim}{1}_{\Theta_{\alpha \underline{i}}} \left(- \frac{1}{\sin \Theta_{\alpha \underline{i}}} \frac{\partial}{\partial \Phi_{\alpha \underline{i}}} + \frac{\cos \Theta_{\alpha \underline{i}}}{\sin \Theta_{\alpha \underline{i}}} \frac{\partial}{\partial \Xi_{\alpha \underline{i}}} \right) \tag{12}$$

is the <u>rotational</u> <u>gradient</u> <u>operator</u> associated with the Eulerian angles $\underset{\sim}{R}_{\alpha \underline{i}} = (\Phi_{\alpha \underline{i}}, \Theta_{\alpha \underline{i}}, \Xi_{\alpha \underline{i}})$ specifying the orientation of the $\underline{i}^{th}$ molecule of species α. To derive this relationship, we imagine the $\underline{i}^{th}$ molecule of species α to undergo a small virtual rigid rotation $d\Psi_{\alpha i} = d\Phi_{\alpha i}\, k + d\Theta_{\alpha i}\, l_{\Phi_{\alpha i}} + d\Xi_{\alpha i}\, l_{R_{\alpha i}}$. The work done against the torque $T_{\alpha i}$ is $- T_{\alpha i} \cdot d\Psi_{\alpha i}$ and this must be equal to the increase in the potential energy which is $dU = \frac{\partial U}{\partial \Phi_{\alpha i}} d\Phi_{\alpha i} + \frac{\partial U}{\partial \Theta_{\alpha i}} d\Theta_{\alpha i} + \frac{\partial U}{\partial \Xi_{\alpha i}} d\Xi_{\alpha i}$. Upon equating coefficients of the arbitrary increments $d\Phi_{\alpha i}$, $d\Theta_{\alpha i}$, and $d\Xi_{\alpha i}$ to zero, we obtain equation (11). Incidently, if $dq_{\alpha ik}$ represents a rotation of the $\underline{i}^{th}$ molecule of species α around some axis, then $F_{\alpha ik}$ is the component of the total applied torque about that axis of rotation. Thus the component of $T_{\alpha i}$ about the axis k is $-\frac{\partial U}{\partial \Phi_{\alpha i}}$; the component about $l_{\Phi_{\alpha i}}$ is $-\frac{\partial U}{\partial \Theta_{\alpha i}}$; and the component about $l_{R_{\alpha i}}$ is $-\frac{\partial U}{\partial \Xi_{\alpha i}}$. With the potential energy of the system given by equation (6), the total torque on molecule $\underline{i}$ of species α becomes

$$T_{\alpha i}(q) = T_{\alpha i}^{(e)}(q_{\alpha i}) + \sum_{\beta=1}^{\sigma} \sum_{\substack{j=1 \\ \beta j \neq \alpha i}}^{N_\beta} T_{\beta j \alpha i}(q_{\beta j}, q_{\alpha i}) \tag{13}$$

where $\underset{\sim}{\underline{T}}_{\alpha \underline{i}}^{(e)}(\underset{\sim}{\underline{q}}_{\alpha \underline{i}}) = -\nabla_{\underset{\sim}{\underline{R}}_{\alpha \underline{i}}} \underline{V}_{\alpha}^{(e)}(\underset{\sim}{\underline{q}}_{\alpha \underline{i}})$ is the torque exerted on molecule $\underline{i}$ of species α by the external field; and where $\underset{\sim}{\underline{T}}_{\beta \underline{j} \alpha \underline{i}}(\underset{\sim}{\underline{q}}_{\beta \underline{j}}, \underset{\sim}{\underline{q}}_{\alpha \underline{i}}) = -\nabla_{\underset{\sim}{\underline{R}}_{\alpha \underline{i}}} \underline{V}_{\alpha\beta}(\underset{\sim}{\underline{q}}_{\alpha \underline{i}}, \underset{\sim}{\underline{q}}_{\beta \underline{j}})$ is the torque exerted on molecule $\underline{i}$ of species α by molecule $\underline{j}$ of species β.

Distribution Functions

The probability density function or the specific distribution function in the $12\underline{N}$-dimensional phase space of the $\underline{N}$ molecule system we denote by $\underline{f}^{(\underline{N})}(\underset{\sim}{\underline{q}}, \underset{\sim}{\underline{p}}; \underline{t})$. Thus $\underline{f}^{(\underline{N})}(\underset{\sim}{\underline{q}}, \underset{\sim}{\underline{p}}; \underline{t})\ d\underset{\sim}{\underline{q}}\ d\underset{\sim}{\underline{p}}$ is the probability that at time $\underline{t}$ the system is in the element of extension $d\underset{\sim}{\underline{q}}\ d\underset{\sim}{\underline{p}}$ at $(\underset{\sim}{\underline{q}}, \underset{\sim}{\underline{p}})$. The boundary conditions upon $\underline{f}^{(\underline{N})}$ require that it vanish near the walls of the container of volume $\underline{V}$ to which the system is confined and for momenta greater than some arbitrary but finite upper bound. Furthermore, we normalize to unity the integral of $\underline{f}^{(N)}$ over all accessible regions of phase space, i.e., we require that

$$\iint \underline{f}^{(\underline{N})}(\underset{\sim}{\underline{q}}, \underset{\sim}{\underline{p}}; \underline{t})\ d\underset{\sim}{\underline{q}}\ d\underset{\sim}{\underline{p}} = 1.$$

In characterizing the macroscopic state of the system, it is convenient to employ lower order specific distribution

<u>functions</u>. For example, the <u>singlet distribution function</u>

$$\underline{f}_{\alpha\underline{i}}^{(1)}(\underset{\sim}{q}_\alpha, \underset{\sim}{p}_\alpha; \underline{t}) = \iint \delta(\underset{\sim}{q}_{\alpha\underline{i}} - \underset{\sim}{q}_\alpha)\ \delta(\underset{\sim}{p}_{\alpha\underline{i}} - \underset{\sim}{p}_\alpha)\ \underline{f}^{(\underline{N})}(\underset{\sim}{q}, \underset{\sim}{p}; \underline{t})\ \underline{d}\underset{\sim}{q}\ \underline{d}\underset{\sim}{p}$$

gives the probability (per unit volume in the 12-dimensional phase space of a single molecule of species α) that at the time $\underline{t}$ molecule $\underline{i}$ of species α will be at position $\underset{\sim}{r}_\alpha$ in orientation $\underset{\sim}{R}_\alpha$ with linear momentum $\underset{\sim}{m}_\alpha$ and with rotational momenta $\underset{\sim}{M}_\alpha$. A further integration yields the singlet distribution function

$$\underline{f}_{\alpha\underline{i}}^{(1)}(\underset{\sim}{r}_\alpha; \underline{t}) = \iint \underline{f}_{\alpha\underline{i}}^{(1)}(\underset{\sim}{q}_\alpha, \underset{\sim}{p}_\alpha; \underline{t})\ \underline{d}^3\underset{\sim}{R}_\alpha\ \underline{d}^6\underset{\sim}{p}_\alpha$$

which gives the probability (per unit volume in the 3-dimensional position space of a single molecule of species α) that molecule $\underline{i}$ of species α will be at position $\underset{\sim}{r}_\alpha$ at time $\underline{t}$. The <u>pair distribution function</u>

$$\underline{f}_{\alpha\underline{i}\beta\underline{j}}^{(2)}(\underset{\sim}{q}_\alpha, \underset{\sim}{p}_\alpha; \underset{\sim}{q}_\beta, \underset{\sim}{p}_\beta; \underline{t})$$

$$= \iint \delta(\underset{\sim}{q}_{\alpha\underline{i}} - \underset{\sim}{q}_\alpha)\ \delta(\underset{\sim}{p}_{\alpha\underline{i}} - \underset{\sim}{p}_\alpha)\ \delta(\underset{\sim}{q}_{\beta\underline{j}} - \underset{\sim}{q}_\beta)\ \delta(\underset{\sim}{p}_{\beta\underline{j}} - \underset{\sim}{p}_\beta)\ \underline{f}^{(\underline{N})}(\underset{\sim}{q}, \underset{\sim}{p}; \underline{t})\ \underline{d}\underset{\sim}{q}\ \underline{d}\underset{\sim}{p}$$

gives the probability (per unit volume in the 24-dimensional phase space of a pair of molecules, one of species α and one of species β) that at time $\underline{t}$ molecule $\underline{i}$ of species α will be at position $\underline{\underset{\sim}{r}}_\alpha$ in orientation $\underline{\underset{\sim}{R}}_\alpha$ moving with linear momentum $\underline{\underset{\sim}{m}}_\alpha$ and rotational momenta $\underline{\underset{\sim}{M}}_\alpha$; and that, at the same time, molecule $\underline{j}$ of species β will be at position $\underline{\underset{\sim}{r}}_\beta$ in orientation $\underline{\underset{\sim}{R}}_\beta$ moving with linear momentum $\underline{\underset{\sim}{m}}_\beta$ and rotational momenta $\underline{\underset{\sim}{M}}_\beta$. Further integration yields the pair distribution function

$$\underline{f}_{\alpha\underline{i}\beta\underline{j}}^{(2)}(\underline{\underset{\sim}{q}}_\alpha;\ \underline{\underset{\sim}{q}}_\beta;\underline{t}) = \iint \underline{f}_{\alpha\underline{i}\beta\underline{j}}^{(2)}(\underline{\underset{\sim}{q}}_\alpha,\underline{\underset{\sim}{p}}_\alpha;\underline{\underset{\sim}{q}}_\beta,\underline{\underset{\sim}{p}}_\beta;\underline{t})\ \underline{d}^6\underline{\underset{\sim}{p}}_\alpha\ \underline{d}^6\underline{\underset{\sim}{p}}_\beta$$

which gives the probability (per unit volume in 12-dimensional $\alpha\beta$-pair configuration space) that at time $\underline{t}$ molecule $\underline{i}$ of species α will be at position $\underline{\underset{\sim}{r}}_\alpha$ in orientation $\underline{\underset{\sim}{R}}_\alpha$ and that, at the same time, molecule $\underline{j}$ of species β will be at position $\underline{\underset{\sim}{r}}_\beta$ in orientation $\underline{\underset{\sim}{R}}_\beta$. Each of these specific distribution functions is normalized to unity.

Usually, we will be primarily concerned with whether or not _any_ molecules of a given species have given positions, orientations, linear momenta, and/or rotational momenta. Accordingly, we introduce _generic distribution functions_. The generic distribution functions defined over singlet and pair subspaces of the $12\underline{N}$-dimensional phase space for the $\underline{N}$ molecule system are related to the specific distribution functions defined over the same subspaces by the equations:

$$\underline{C}_{\alpha}^{(1)} = \sum_{\underline{i}=1}^{\underline{N}_{\alpha}} \underline{f}_{\alpha\underline{i}}^{(1)} = \underline{N}_{\alpha}\ \underline{f}_{\alpha\underline{i}}^{(1)}$$

$$\underline{C}_{\alpha\beta}^{(2)} = \sum_{\underline{i}=1}^{\underline{N}_{\alpha}} \sum_{\substack{\underline{j}=1 \\ \alpha\underline{i} \neq \beta\underline{j}}}^{\underline{N}_{\beta}} \underline{f}_{\alpha\underline{i}\beta\underline{j}}^{(2)} = \underline{N}_{\alpha}(\underline{N}_{\beta} - \delta_{\alpha\beta})\ \underline{f}_{\alpha\underline{i}\beta\underline{j}}^{(2)}.$$

For example, the <u>generic singlet distribution function</u> $\underline{C}_{\alpha}^{(1)}(\underline{\underline{r}}_{\alpha};\underline{t}) = \underline{N}_{\alpha}\ \underline{f}_{\alpha\underline{i}}^{(1)}(\underline{\underline{r}}_{\alpha};\underline{t})$ gives the probability (per unit volume in α-position space) that, at time $\underline{t}$, a molecule of species α will be at position $\underline{\underline{r}}_{\alpha}$. Similarly, $\underline{C}_{\alpha}^{(1)}(\underline{\underline{q}}_{\alpha};\underline{t}) = \underline{N}_{\alpha}\ \underline{f}_{\alpha\underline{i}}^{(1)}(\underline{\underline{q}}_{\alpha};\underline{t})$ gives the probability (per unit volume in α-configuration space) that, at time $\underline{t}$, a molecule of species α will be at position $\underline{\underline{r}}_{\alpha}$ in orientation $\underline{\underline{R}}_{\alpha}$. The <u>generic pair distribution function</u>

$$\underline{C}_{\alpha\beta}^{(2)}(\underline{\underline{q}}_{\alpha};\underline{\underline{q}}_{\beta};\underline{t}) = \underline{N}_{\alpha}(\underline{N}_{\beta}-\delta_{\alpha\beta})\ \underline{f}_{\alpha\underline{i}\beta\underline{j}}^{(2)}(\underline{\underline{q}}_{\alpha};\underline{\underline{q}}_{\beta};\underline{t})$$

gives the probability (per unit volume in $\alpha\beta$-pair configuration space) that, at time $\underline{t}$, a molecule of species α will be at position $\underline{\underline{r}}_{\alpha}$ in orientation $\underline{\underline{R}}_{\alpha}$; and that at the same time a molecule of species β will be at position $\underline{\underline{r}}_{\beta}$ in orientation $\underset{\sim}{R}_\beta$. Clearly, the generic distribution functions $c_\alpha^{(1)}$ and $c_{\alpha\beta}^{(2)}$ are normalized to N_α and to $N_\alpha(N_\beta - \delta_{\alpha\beta})$, respectively.

For some purposes it is convenient to use unsymmetrical distribution functions in which the coordinates and momenta of certain molecules play a special role. Thus we introduce <u>conditional specific distribution functions</u> such as

$$f_\alpha^{(1)}(\underset{\sim}{m}_\alpha | \underset{\sim}{r}_\alpha; t) = \frac{f_{\alpha i}^{(1)}(\underset{\sim}{r}_\alpha, \underset{\sim}{m}_\alpha; t)}{f_{\alpha i}^{(1)}(\underset{\sim}{r}_\alpha; t)} = \frac{c_\alpha^{(1)}(\underset{\sim}{r}_\alpha, \underset{\sim}{m}_\alpha; t)}{c_\alpha^{(1)}(\underset{\sim}{r}_\alpha; t)}$$

which (provided $f_{\alpha i}^{(1)}(\underset{\sim}{r}_\alpha; t) \neq 0$) gives the probability (per unit volume in α- linear momentum space) that, at time t, a specific molecule of species α at position $\underset{\sim}{r}_\alpha$ will have linear momentum $\underset{\sim}{m}_\alpha$. The <u>conditional generic distribution function</u>

$$c_{\beta|\alpha}^{(2)}(\underset{\sim}{q}_\beta | \underset{\sim}{q}_\alpha; t) = \frac{c_{\alpha\beta}^{(2)}(\underset{\sim}{q}_\alpha; \underset{\sim}{q}_\beta; t)}{c_\alpha^{(1)}(\underset{\sim}{q}_\alpha; t)} = (N_\beta - \delta_{\alpha\beta}) \frac{f_{\alpha i \beta j}^{(2)}(\underset{\sim}{q}_\alpha; \underset{\sim}{q}_\beta; t)}{f_{\alpha i}^{(1)}(\underset{\sim}{q}_\alpha; t)}$$

(provided $f_{\alpha i}^{(1)}(\underset{\sim}{q}_\alpha; t) \neq 0$) gives the probability (per unit volume in β- configuration space) of observing a molecule of species β at position $\underset{\sim}{r}_\beta$ in orientation $\underset{\sim}{R}_\beta$, at time t, relative to the hypothesis that a molecule of species α

is at position $\underset{\sim}{r}_\alpha$ and in orientation $\underset{\sim}{R}_\alpha$.

If the relationship $\underline{f}_{\alpha \underline{i} \beta \underline{j}}^{(2)}(\underset{\sim}{q}_\alpha; \underset{\sim}{q}_\beta; \underline{t})$ = $\underline{f}_{\alpha \underline{i}}^{(1)}(\underset{\sim}{q}_\alpha; \underline{t})\ \underline{f}_{\beta \underline{j}}^{(1)}(\underset{\sim}{q}_\beta; \underline{t})$ holds, it follows that $\underline{C}_{\beta|\alpha}^{(2)}(\underset{\sim}{q}_\beta | \underset{\sim}{q}_\alpha; \underline{t}) = ((\underline{N}_\beta - \delta_{\alpha\beta})/\underline{N}_\beta)\ \underline{C}_\beta^{(1)}(\underset{\sim}{q}_\beta; \underline{t})$ and that

$$\underline{C}_{\alpha\beta}^{(2)}(\underset{\sim}{q}_\alpha; \underset{\sim}{q}_\beta; \underline{t}) = \frac{(\underline{N}_\beta - \delta_{\alpha\beta})}{\underline{N}_\beta}\ \underline{C}_\alpha^{(1)}(\underset{\sim}{q}_\alpha; \underline{t})\ \underline{C}_\beta^{(1)}(\underset{\sim}{q}_\beta; \underline{t}). \quad (14)$$

In this case (provided that $\alpha \neq \beta$; or, if $\alpha = \beta$, provided that $\underline{N}_\alpha \gg 1$) the probability (per unit volume in β-configuration space) of a molecule of species β being in position $\underset{\sim}{r}_\beta$ and in orientation $\underset{\sim}{R}_\beta$ is independent of whether or not a molecule of species α is at position $\underset{\sim}{r}_\alpha$ and in orientation $\underset{\sim}{R}_\alpha$. For a fluid we judge these two events to be causally independent, and thus assume equation (14) to be valid, when the length $\underline{r}_{\beta\alpha}$ of the vector $\underset{\sim}{r}_{\beta\alpha}$ = $\underset{\sim}{r}_\beta - \underset{\sim}{r}_\alpha$ is sufficiently large so as to exceed the ranges of the intermolecular forces. This suggests that we introduce an $\alpha\beta$-<u>pair</u> <u>configuration</u> <u>correlation</u> <u>function</u> $\underline{G}_{\alpha\beta}^{(2)}(\underset{\sim}{q}_\alpha; \underset{\sim}{q}_\beta; \underline{t})$ which represents the factor by which $\underline{C}_{\alpha\beta}^{(2)}(\underset{\sim}{q}_\alpha; \underset{\sim}{q}_\beta; \underline{t})$ deviates from the "causally independent" value given by equation (14). We define

$$\underline{C}_{\alpha\beta}^{(2)}(\underset{\sim}{\underline{q}}_\alpha;\underset{\sim}{\underline{q}}_\beta;\underline{t}) = \underline{C}_\alpha^{(1)}(\underset{\sim}{\underline{q}}_\alpha;\underline{t})\ \underline{C}_\beta^{(1)}(\underset{\sim}{\underline{q}}_\beta;\underline{t})\ \underline{G}_{\alpha\beta}^{(2)}(\underset{\sim}{\underline{q}}_\alpha;\underset{\sim}{\underline{q}}_\beta;\underline{t}). \quad (15)$$

Thus for a fluid, $\underline{G}_{\alpha\beta}^{(2)}(\underset{\sim}{\underline{q}}_\alpha;\ \underset{\sim}{\underline{q}}_\beta;\underline{t})$, which is a function of $\underset{\sim}{\underline{r}}_\alpha$, $\underset{\sim}{\underline{R}}_\alpha$, $\underset{\sim}{\underline{R}}_\beta$, $\underset{\sim}{\underline{r}}_{\beta\alpha}$, and $\underline{t}$, approaches $1 - \delta_{\alpha\beta}/\underline{N}_\alpha$ as $\underline{r}_{\beta\alpha}$ approaches infinity, i.e.,

$$\lim_{\underline{r}_{\alpha\beta}\to+\infty} \underline{G}_{\alpha\beta}^{(2)}(\underset{\sim}{\underline{q}}_\alpha;\ \underset{\sim}{\underline{q}}_\beta;\underline{t}) = 1 - \frac{\delta_{\alpha\beta}}{\underline{N}_\alpha}\ ; \quad (16)$$

and deviations from $1 - \delta_{\alpha\beta}/\underline{N}_\alpha$ are a measure of correlations in the positions and orientations of pairs of molecules.

Because of the repulsive forces between molecules separated by small intermolecular distances, we have

$$\lim_{\underline{r}_{\alpha\beta}\to 0} \underline{G}_{\alpha\beta}^{(2)}(\underset{\sim}{\underline{q}}_\alpha;\ \underset{\sim}{\underline{q}}_\beta;\underline{t}) = 0\ . \quad (17)$$

If the molecules are considered to be convex rigid bodies, this last boundary condition upon $\underline{G}_{\alpha\beta}^{(2)}(\underset{\sim}{\underline{q}}_\alpha;\underset{\sim}{\underline{q}}_\beta;\underline{t})$ can be expressed in a different way. We introduce a <u>closest approach parameter</u> $\underline{a}_{\alpha\beta}$ or $\underline{a}_{\alpha\beta}(\underset{\sim}{\underline{R}}_\alpha, \underset{\sim}{\underline{R}}_\beta; \underset{\sim}{\underline{1}}\underline{r}_{\beta\alpha})$ defined as the minimum distance of approach of the center of mass $\underset{\sim}{\underline{r}}_\alpha$ of a molecule of species α in orientation $\underset{\sim}{\underline{R}}_\alpha$ to the center of mass $\underset{\sim}{\underline{r}}_\beta$ of a molecule of species β in orientation $\underset{\sim}{\underline{R}}_\beta$, the direction of approach being characterized by the unit vector $\underset{\sim}{1}\underline{r}_{\beta\alpha} = (1/\underline{r}_{\beta\alpha})\,\underset{\sim}{\underline{r}}_{\beta\alpha}$ directed from the center of mass of the molecule of species α to the center of mass of the molecule of species β. Then the hydrodynamical requirement that the convex rigid molecules do not interpenetrate one another can be written as

$$\underline{G}_{\alpha\beta}^{(2)}(\underset{\sim}{\underline{q}}_{\alpha};\, \underset{\sim}{\underline{q}}_{\beta};\, \underline{t}) = 0 \quad \text{if} \quad \underline{r}_{\alpha\beta} < \underline{a}_{\alpha\beta}(\underset{\sim}{\underline{R}}_{\alpha}, \underset{\sim}{\underline{R}}_{\beta};\, \underset{\sim}{1}\underline{r}_{\beta\alpha}). \qquad (18)$$

For molecules assumed to be rigid spheres the closest approach parameter $\underline{a}_{\alpha\beta}$ is a constant independent of molecular orientations and the direction of approach.

Finally, we note that the $\alpha\beta$-<u>pair position correlation function</u> $\underline{G}_{\alpha\beta}^{(2)}(\underset{\sim}{\underline{r}}_{\alpha};\, \underset{\sim}{\underline{r}}_{\beta};\, \underline{t})$ defined by

$$\underline{c}_{\alpha\beta}^{(2)}(\underset{\sim}{\underline{r}}_{\alpha}; \underset{\sim}{\underline{r}}_{\beta};\, \underline{t}) = \underline{c}_{\alpha}^{(1)}(\underset{\sim}{\underline{r}}_{\alpha};\, \underline{t})\; \underline{c}_{\beta}^{(1)}(\underset{\sim}{\underline{r}}_{\beta};\, \underline{t})\; \underline{G}_{\alpha\beta}^{(2)}(\underset{\sim}{\underline{r}}_{\alpha}; \underset{\sim}{\underline{r}}_{\beta};\, \underline{t}) \qquad (19)$$

is the familiar radial distribution function obtained experimentally by X-ray scattering. The boundary conditions on $\underline{G}_{\alpha\beta}^{(2)}(\underset{\sim}{\underline{r}}_{\alpha};\, \underset{\sim}{\underline{r}}_{\beta};\, \underline{t})$ are the same as those on $\underline{G}_{\alpha\beta}^{(2)}(\underset{\sim}{\underline{q}}_{\alpha}; \underset{\sim}{\underline{q}}_{\beta};\, \underline{t})$.

Our principal task, now, is to show how the various distribution functions can be calculated theoretically, and how the various macroscopic observables are related to them.

Although we shall employ certain distribution functions which have not been expressly defined here, the notations and their meanings are obvious.

Statistical Mechanical Expressions for Macroscopic Observables

The fundamental assumption of this approach to statistical mechanics is that: given the probability density function $\underline{f}^{(\underline{N})}(\underset{\sim}{\underline{q}}, \underset{\sim}{\underline{p}}; \underline{t})$, the macroscopic observable $\underline{G}_{\underline{obs}}$ corresponding to any dynamical variable $\underline{G}(\underset{\sim}{\underline{q}}, \underset{\sim}{\underline{p}})$ of the system can be identified with the average value of the dynamical variable over the density, i.e.,

$$\underline{G}_{\underline{obs}} = \overline{\underline{G}} = \langle \underline{G} \rangle = \left\langle \underline{G};\ \underline{f}^{(\underline{N})} \right\rangle = \iint \underline{G}(\underset{\sim}{\underline{q}},\underset{\sim}{\underline{p}})\ \underline{f}^{(\underline{N})}(\underset{\sim}{\underline{q}}, \underset{\sim}{\underline{p}}; \underline{t})\ d\underset{\sim}{\underline{q}}\ d\underset{\sim}{\underline{p}}. \quad (20)$$

We shall now identify some of the important macroscopic observables with averages over $\underline{f}^{(\underline{N})}(\underset{\sim}{\underline{q}}, \underset{\sim}{\underline{p}}; \underline{t})$, or rather with averages over appropriate lower order distribution functions.

The composition of the fluid in the element of volume $d^3\underset{\sim}{\underline{r}}$ about $\underset{\sim}{\underline{r}}$, at a given time $\underline{t}$, is given by the <u>number densities</u> of the σ species

$$\underline{c}_\alpha^{(1)}(\underset{\sim}{\underline{r}}; \underline{t}) = \lim_{\Delta \underset{\sim}{\underline{r}} \to 0} \frac{\Delta \underline{N}_\alpha(\underset{\sim}{\underline{r}}; \underline{t})}{\Delta \underset{\sim}{\underline{r}}\ (\underset{\sim}{\underline{r}}; \underline{t})}$$

where $\alpha = 1,\ldots,\sigma$; and where $\Delta \underline{N}_\alpha(\underline{\underline{r}}; \underline{t})$ is the number of α molecules or ions in the element of volume $\Delta \underline{\underline{r}}\ (\underline{\underline{r}}; \underline{t})$ about $\underline{\underline{r}}$ at time $\underline{t}$. We identify these number densities with the singlet generic distribution functions for which we have used the same notations. Alternately, the composition of the fluid can be specified by the partial mass densities $\rho_\alpha(\underline{\underline{r}}; \underline{t}) = \underline{m}_\alpha\ \underline{C}_\alpha^{(1)}(\underline{\underline{r}}; \underline{t})$ where $\alpha = 1,\ldots,\sigma$. The total mass density of the fluid is given by $\rho(\underline{\underline{r}}; \underline{t}) = \sum_{\alpha=1}^{\sigma} \rho_\alpha(\underline{\underline{r}}; \underline{t})$.

In order to describe the currents of matter within the fluid, we define mean local velocities for the various species present. The mean local velocity, at position $\underline{\underline{r}}$ and at time $\underline{t}$, of a set of molecules all of the same species, is the average velocity of those molecules of the set found in the element of volume $\underline{d}^3\underline{\underline{r}}$ about $\underline{\underline{r}}$ at time $\underline{t}$. We identify this with the average velocity, at time $\underline{t}$, of a molecule of the set provided it is at position $\underline{\underline{r}}$. Thus the mean local velocity at position $\underline{\underline{r}}_\alpha$, and at time $\underline{t}$, of the molecules of species α in the particular orientation $\underline{\underline{R}}_\alpha$ is

$$\underline{\underline{u}}_\alpha(\underline{\underline{q}}_\alpha; \underline{t}) = \frac{1}{\underline{m}_\alpha} \int \underline{\underline{m}}_\alpha\ \underline{f}_\alpha^{(1)}(\underline{\underline{m}}_\alpha \mid \underline{\underline{q}}_\alpha; \underline{t})\ \underline{d}^3\underline{\underline{m}}_\alpha .$$

Similarly, the over-all mean local velocity at position $\underline{\underline{r}}_\alpha$, and at time t, of the molecules of species α is

$$\underset{\sim}{u}_\alpha(\underset{\sim}{r}_\alpha; t) = \frac{1}{m_\alpha} \int \underset{\sim}{m}_\alpha \, f_\alpha^{(1)}(\underset{\sim}{m}_\alpha | \underset{\sim}{r}_\alpha; t) \, d^3\underset{\sim}{m}_\alpha$$

$$= \int \underset{\sim}{u}_\alpha(\underset{\sim}{q}_\alpha; t) \, f_\alpha^{(1)}(\underset{\sim}{R}_\alpha | \underset{\sim}{r}_\alpha; t) \, d^3\underset{\sim}{R}_\alpha .$$

The <u>particle current density</u> of species α at position $\underset{\sim}{r}_\alpha$ and at time t, i.e., the rate at which molecules or ions of species α pass through a unit element of surface at the position $\underset{\sim}{r}_\alpha$ at time t, is $\underset{\sim}{J}_\alpha(\underset{\sim}{r}_\alpha; t) = C_\alpha^{(1)}(\underset{\sim}{r}_\alpha; t) \, \underset{\sim}{u}_\alpha(\underset{\sim}{r}_\alpha; t)$; and the particle current density of the molecules of species α in the particular orientation $\underset{\sim}{R}_\alpha$, again at position $\underset{\sim}{r}_\alpha$ and at time t, is $\underset{\sim}{J}_\alpha(\underset{\sim}{q}_\alpha; t) = C_\alpha^{(1)}(\underset{\sim}{q}_\alpha; t) \, \underset{\sim}{u}_\alpha(\underset{\sim}{q}_\alpha; t)$. The <u>mass current density</u> or the mass flux of species α at position $\underset{\sim}{r}_\alpha$ and at time t is

$$\rho_\alpha(\underset{\sim}{r}_\alpha; t) \, \underset{\sim}{u}_\alpha(\underset{\sim}{r}_\alpha; t) = \int \underset{\sim}{m}_\alpha \, C_\alpha^{(1)}(\underset{\sim}{r}_\alpha, \underset{\sim}{m}_\alpha; t) \, d^3\underset{\sim}{m}_\alpha .$$

Upon adding contributions from all the species present, we obtain the <u>total mass current density</u> at position $\underset{\sim}{r}$ and at time t:

$$\rho(\underset{\sim}{r}; t) \, \underset{\sim}{u}(\underset{\sim}{r}; t) = \sum_{\alpha=1}^{\sigma} \rho_\alpha(\underset{\sim}{r}; t) \, \underset{\sim}{u}_\alpha(\underset{\sim}{r}; t) .$$

This equation serves to define $\underset{\sim}{u}(\underset{\sim}{r};\ t)$, the over-all mean local velocity or the macroscopic stream velocity of the fluid at position $\underset{\sim}{r}$ and at time t.

We also introduce relative particle current densities, or particle diffusion currents, defined by

$$\underset{\sim}{J}_{\alpha}^{(rel)}(\underset{\sim}{r};\ t) = C_{\alpha}^{(1)}(\underset{\sim}{r};\ t)\left[\underset{\sim}{u}_{\alpha}(\underset{\sim}{r};\ t) - \underset{\sim}{u}(\underset{\sim}{r};\ t)\right] \tag{21}$$

where $\alpha = 1, \ldots, \sigma$. These diffusion currents are defined relative to the motion of the fluid as a whole. Often, it is convenient to relate the particle diffusion currents to the motion of the solvent — species τ. We define

$$\underset{\sim}{J}_{\alpha}^{(rel\ \tau)}(\underset{\sim}{r};\ t) = C_{\alpha}^{(1)}(\underset{\sim}{r};\ t)\left[\underset{\sim}{u}_{\alpha}(\underset{\sim}{r};\ t) - \underset{\sim}{u}_{\tau}(\underset{\sim}{r};\ t)\right] \tag{22}$$

where $\alpha = 1, \ldots, \sigma$. These diffusion currents are related to the former ones by:

$$\underset{\sim}{J}_{\alpha}^{(rel\ \tau)}(\underset{\sim}{r};\ t) = \underset{\sim}{J}_{\alpha}^{(rel)}(\underset{\sim}{r};\ t) - \frac{C_{\alpha}^{(1)}(\underset{\sim}{r};\ t)}{C_{\tau}^{(1)}(\underset{\sim}{r};\ t)}\underset{\sim}{J}_{\tau}^{(rel)}(\underset{\sim}{r};\ t) \tag{23}$$

where $\alpha = 1, \ldots, \sigma$. For dilute solutions, if $\alpha \neq \tau$, then $C_{\alpha}^{(1)}(\underset{\sim}{r};\ t) / C_{\tau}^{(1)}(\underset{\sim}{r};\ t) \cong 0$ whence $\underset{\sim}{J}_{\alpha}^{(rel\ \tau)}(\underset{\sim}{r};\ t) \cong$

$\underset{\sim}{J}_\alpha^{(rel)}(\underset{\sim}{r}; t)$.

The partial electric current density $\underset{\sim}{I}_\alpha(\underset{\sim}{r}; t) = e\, Z_\alpha\, \underset{\sim}{J}_\alpha^{(rel\, T)}(\underset{\sim}{r}; t)$ is the rate at which electric charge passes through a unit element of surface at the position $\underset{\sim}{r}$ and at time t due to the diffusion of species α. Here e is the protonic charge and Z_α is the valence of species α. The total electric current density is:

$$\underset{\sim}{I}(\underset{\sim}{r}; t) = \sum_{\alpha=1}^{\sigma} \underset{\sim}{I}_\alpha(\underset{\sim}{r}; t) = e \sum_{\alpha=1}^{\sigma} Z_\alpha\, \underset{\sim}{J}_\alpha^{(rel\, T)}(\underset{\sim}{r}; t) \tag{24}$$

$$= e \sum_{\alpha=1}^{\sigma} Z_\alpha\, \underset{\sim}{J}_\alpha^{(rel)}(\underset{\sim}{r}; t) = e \sum_{\alpha=1}^{\sigma} Z_\alpha\, \underset{\sim}{J}_\alpha(\underset{\sim}{r}; t) .$$

The last two equalities follow from equations (22) and (23) and from the condition for local electroneutrality

$$\sum_{\alpha=1}^{\sigma} Z_\alpha\, c_\alpha^{(1)}(\underset{\sim}{r}; t) = 0 .$$

The transference number t_α of species α is the fraction of the total current[19] which is carried by species α, i.e.,

$$t_\alpha(\underset{\sim}{r}; t) = \frac{\underset{\sim}{I}_\alpha(\underset{\sim}{r}; t)}{\underset{\sim}{I}(\underset{\sim}{r}; t)} .$$

Since the valence of a (neutral) non-electrolyte species is zero, its transference number vanishes. Therefore, in order to describe the currents of neutral species accompanying the transport of ionic species carrying an electric current we introduce the mass transport ratio,

$$\underline{t}_{\alpha}^{(m)}(\underline{r};\ \underline{t}) = \frac{\underline{m}_{\alpha}\ \underline{J}_{\alpha}^{(\mathrm{rel}\ \tau)}(\underline{r};\ \underline{t})}{\underline{I}(\underline{r};\ \underline{t})} \tag{25}$$

equal to the mass current of species α (relative to the solvent τ) associated with the passage of unit electric current.

The particle current density, in $\alpha\beta$- pair space, of molecules of species α in orientation $\underline{R}_{\alpha}$ and of molecules of species β in orientation $\underline{R}_{\beta}$ is the six-component vector

$$\underline{J}_{\alpha\beta}^{(2)}(\underline{q}_{\alpha};\ \underline{q}_{\beta};\ \underline{t}) = \underline{J}_{\alpha\beta,1}^{(2)}(\underline{q}_{\alpha};\underline{q}_{\beta};\ \underline{t}) \oplus \underline{J}_{\alpha\beta,2}^{(2)}(\underline{q}_{\alpha};\underline{q}_{\beta};\ \underline{t})$$

$$= \iint \left(\frac{\underline{m}_{\alpha}}{m_{\alpha}} \oplus \frac{\underline{m}_{\beta}}{m_{\beta}} \right) \underline{C}_{\alpha\beta}^{(2)}(\underline{q}_{\alpha},\ \underline{m}_{\alpha};\ \underline{q}_{\beta},\ \underline{m}_{\beta};\ \underline{t})\ \underline{d}^3\underline{m}_{\alpha}\ \underline{d}^3\underline{m}_{\beta}\ .$$

Also, $\underline{J}_{\alpha\beta}^{(2)}(\underline{q}_{\alpha};\underline{q}_{\beta};\ \underline{t}) = \underline{C}_{\alpha\beta}^{(2)}(\underline{q}_{\alpha};\underline{q}_{\beta};\ \underline{t})\ \ \underline{u}_{\alpha\beta}^{(2)}(\underline{q}_{\alpha};\underline{q}_{\beta};\ \underline{t})$

where $\underline{u}_{\alpha\beta}^{(2)}(\underline{q}_{\alpha};\underline{q}_{\beta};\ \underline{t}) = \underline{u}_{\alpha\beta,1}^{(2)}(\underline{q}_{\alpha};\underline{q}_{\beta};\underline{t}) \oplus \underline{u}_{\alpha\beta,2}^{(2)}(\underline{q}_{\alpha};\underline{q}_{\beta};\underline{t})$

is the mean local velocity in $\alpha\beta$- pair space. The three-component vector

$$\underset{\sim}{u}_{\alpha\beta,1}^{(2)}(\underset{\sim}{q}_\alpha;\underset{\sim}{q}_\beta;\underline{t}) = \frac{1}{\underline{n}_\alpha}\int \underset{\sim}{m}_\alpha\ \underline{f}_{\alpha|\alpha\beta}^{(2)}(\underset{\sim}{m}_\alpha|\underset{\sim}{q}_\alpha;\underset{\sim}{q}_\beta;\underline{t})\ \underline{d}^3\underset{\sim}{m}_\alpha$$

is the mean local velocity at position $\underset{\sim}{r}_\alpha$, and at time $\underline{t}$, of the molecules of species α in the particular orientation $\underset{\sim}{R}_\alpha$, provided a molecule of species β is at position $\underset{\sim}{r}_\beta$ in orientation $\underset{\sim}{R}_\beta$. Definitions similar to these can of course be given for $\underset{\sim}{J}_{\alpha\beta}^{(2)}(\underset{\sim}{r}_\alpha;\underset{\sim}{r}_\beta;\underline{t})$ and $\underset{\sim}{u}_{\alpha\beta}^{(2)}(\underset{\sim}{r}_\alpha;\underset{\sim}{r}_\beta;\underline{t})$.

The hydrodynamical requirement (see equations 17 and 18) that two molecules schematically represented as convex rigid bodies do not interpenetrate one another is tantamount to the requirement that when the two molecules are in contact (in fixed orientations with a constant unit vector being directed from the center of mass of one of the molecules toward the center of mass of the other) the component of their relative velocity along the unit vector directed between their mass centers must vanish, i.e.,

$$\left[\underset{\sim}{J}_{\alpha\beta,1}^{(2)}(\underset{\sim}{q}_\alpha;\underset{\sim}{q}_\beta;\underline{t}) - \underset{\sim}{J}_{\alpha\beta,2}^{(2)}(\underset{\sim}{q}_\alpha;\underset{\sim}{q}_\beta;\underline{t})\right]_{\underline{r}_{\alpha\beta}=\underline{a}_{\alpha\beta}} \cdot \underset{\sim}{1}_{\underline{r}_{\beta\alpha}} = 0. \quad (26)$$

As noted before, if the molecules are assumed to be rigid

spheres the closest approach parameter $a_{\alpha\beta}(R_\alpha, R_\beta; {}^1r_{\beta\alpha})$ is a constant characteristic of the two species α and β.

Having previously introduced mean local translational velocities, we now define mean local angular velocities. The mean local angular velocity, at position r and at time t, of a set of molecules all of the same species and in the same orientation, is the average angular velocity of those molecules of the set found in the element of volume d^3r about r at time t. We identify this with the average angular velocity, at time t, of a molecule of the set provided it is at the particular position r. Thus

$$\Omega_\alpha(q_\alpha; t) = \int \omega_\alpha \, f_\alpha^{(1)}(M_\alpha \mid q_\alpha; t) \; d^3M_\alpha$$

is the mean local angular velocity at position r_α, and at time t, of the molecules of species α in orientation R_α. The average angular momentum (associated with the rotations of the molecules themselves) of the molecules of species α in the orientation R_α found in the element of volume d^3r_α about r_α at time t is

$$\Lambda_\alpha(q_\alpha; t) = \int L_\alpha \; f_\alpha^{(1)}(M_\alpha \mid q_\alpha; t) \; d^3M_\alpha$$

and the average angular momentum of all the molecules of species α found in the element of volume $\underline{d}^3\underline{\underset{\sim}{r}}_\alpha$ about $\underline{\underset{\sim}{r}}_\alpha$ at time $\underline{t}$ is

$$\underset{\sim}{\Lambda}_\alpha(\underline{\underset{\sim}{r}}_\alpha;\underline{t}) = \int \underset{\sim}{\Lambda}_\alpha(\underline{\underset{\sim}{q}}_\alpha;\underline{t})\ \underline{f}_\alpha^{(1)}(\underline{\underset{\sim}{R}}_\alpha \mid \underline{\underset{\sim}{r}}_\alpha;\underline{t})\ \underline{d}^3\underline{\underset{\sim}{R}}_\alpha .$$

The average moment of inertia of those molecules of species α found in the element of volume $\underline{d}^3\underline{\underset{\sim}{r}}_\alpha$ about $\underline{\underset{\sim}{r}}_\alpha$ at time $\underline{t}$ is

$$\left\langle\!\left\langle \underline{\underset{\sim}{I}}_\alpha(\underline{\underset{\sim}{r}}_\alpha;\underline{t}) \right\rangle\!\right\rangle = \int \underline{\underset{\sim}{I}}_\alpha\ \underline{f}_\alpha^{(1)}(\underline{\underset{\sim}{R}}_\alpha \mid \underline{\underset{\sim}{r}}_\alpha;\underline{t})\ \underline{d}^3\underline{\underset{\sim}{R}}_\alpha .$$

If it happens that $\left\langle\!\left\langle \underline{\underset{\sim}{I}}_\alpha(\underline{\underset{\sim}{r}}_\alpha;\underline{t}) \right\rangle\!\right\rangle$ is nonsingular and if we write its inverse as $\left(\left\langle\!\left\langle \underline{\underset{\sim}{I}}_\alpha(\underline{\underset{\sim}{r}}_\alpha;\underline{t}) \right\rangle\!\right\rangle \right)^{-1}$, we can define the mean local angular velocity at position $\underline{\underset{\sim}{r}}_\alpha$ and at time $\underline{t}$ of the molecules of species α to be[20] $\underset{\sim}{\Omega}_\alpha(\underline{\underset{\sim}{r}}_\alpha;\underline{t}) = \left(\left\langle\!\left\langle \underline{\underset{\sim}{I}}_\alpha(\underline{\underset{\sim}{r}}_\alpha;\underline{t}) \right\rangle\!\right\rangle \right)^{-1} \cdot \underset{\sim}{\Lambda}_\alpha(\underline{\underset{\sim}{r}}_\alpha;\underline{t})$.

The <u>mean local angular velocity, in $\alpha\beta$- pair space</u>, of molecules of species α in orientation $\underline{\underset{\sim}{R}}_\alpha$ and of molecules of species β in orientation $\underline{\underset{\sim}{R}}_\beta$ is the six-component vector

$$\underset{\sim}{\Omega}_{\alpha\beta}^{(2)}(\underline{\underset{\sim}{q}}_\alpha;\underline{\underset{\sim}{q}}_\beta;\underline{t}) = \underset{\sim}{\Omega}_{\alpha\beta,1}^{(2)}(\underline{\underset{\sim}{q}}_\alpha;\underline{\underset{\sim}{q}}_\beta;\underline{t}) \oplus \underset{\sim}{\Omega}_{\alpha\beta,2}^{(2)}(\underline{\underset{\sim}{q}}_\alpha;\underline{\underset{\sim}{q}}_\beta;\underline{t})$$

defined by

$$\underline{C}_{\alpha\beta}^{(2)}(\underset{\sim}{\underline{q}}_\alpha;\underset{\sim}{\underline{q}}_\beta;\underline{t})\ \underset{\sim}{\Omega}_{\alpha\beta}^{(2)}(\underset{\sim}{\underline{q}}_\alpha;\underset{\sim}{\underline{q}}_\beta;\underline{t}) =$$

$$= \iint (\underset{\sim}{\omega}_\alpha \oplus \underset{\sim}{\omega}_\beta)\ \underline{C}_{\alpha\beta}^{(2)}(\underset{\sim}{\underline{q}}_\alpha,\underset{\sim}{\underline{M}}_\alpha;\ \underset{\sim}{\underline{q}}_\beta,\underset{\sim}{\underline{M}}_\beta;\underline{t})\ \ \underline{d}^3\underset{\sim}{\underline{M}}_\alpha\ \underline{d}^3\underset{\sim}{\underline{M}}_\beta .$$

Clearly,

$$\underset{\sim}{\Omega}_{\alpha\beta,1}^{(2)}(\underset{\sim}{\underline{q}}_\alpha;\underset{\sim}{\underline{q}}_\beta;\underline{t}) = \int \underset{\sim}{\omega}_\alpha\ \underline{f}_{\alpha|\alpha\beta}^{(2)}(\underset{\sim}{\underline{M}}_\alpha \mid \underset{\sim}{\underline{q}}_\alpha;\ \underset{\sim}{\underline{q}}_\beta;\underline{t})\ \underline{d}^3\underset{\sim}{\underline{M}}_\alpha$$

is the mean local angular velocity at position $\underset{\sim}{\underline{r}}_\alpha$ and at time $\underline{t}$ of the molecules of species α in the particular orientation $\underset{\sim}{\underline{R}}_\alpha$, provided a molecule of species β is at position $\underset{\sim}{\underline{r}}_\beta$ and in orientation $\underset{\sim}{\underline{R}}_\beta$.

The <u>mean</u> <u>generalized</u> <u>force</u> associated with $\underline{q}_{\alpha\underline{k}}$, the $\underline{k}$th generalized coordinate used in specifying the configuration $\underset{\sim}{\underline{q}}_\alpha$ of a molecule of species α, averaged over all configurations of the remaining $\underline{N}-1$ molecules of the system is given by

$$\underline{F}_{\alpha\underline{k}}^{(1)}(\underset{\sim}{\underline{q}}_\alpha;\underline{t}) = \frac{\int \underline{F}_{\alpha\underline{i}\underline{k}}\ \delta(\underset{\sim}{\underline{q}}_{\alpha\underline{i}} - \underset{\sim}{\underline{q}}_\alpha)\ \underline{f}^{(\underline{N})}(\underset{\sim}{\underline{q}};\underline{t})\ \ \underline{d}\underset{\sim}{\underline{q}}}{\int \delta(\underset{\sim}{\underline{q}}_{\alpha\underline{i}} - \underset{\sim}{\underline{q}}_\alpha)\ \underline{f}^{(\underline{N})}(\underset{\sim}{\underline{q}};\underline{t})\ \underline{d}\underset{\sim}{\underline{q}}} .$$

Consequently, the total force acting at the center of mass of a molecule of species α in configuration $\underline{q}_\alpha$, averaged over all configurations of the $\underline{N}-1$ remaining molecules of the system, is

$$\underline{F}_\alpha^{(1)}(\underline{q}_\alpha;\underline{t}) = \frac{\int \underline{F}_{\alpha\underline{i}}\ \delta(\underline{q}_{\alpha\underline{i}}-\underline{q}_\alpha)\ \underline{f}^{(\underline{N})}(\underline{q};\underline{t})\ \underline{d}\underline{q}}{\int \delta(\underline{q}_{\alpha\underline{i}}-\underline{q}_\alpha)\ \underline{f}^{(\underline{N})}(\underline{q};\underline{t})\ \underline{d}\underline{q}} .$$

With the molecular interactions assumed to be pair-additive, i.e., with the total force $\underline{F}_{\alpha\underline{i}}$ given by equation (10), this becomes

$$\underline{F}_\alpha^{(1)}(\underline{q}_\alpha;\underline{t}) = \underline{F}_\alpha^{(e)}(\underline{q}_\alpha;\underline{t}) + \sum_{\beta=1}^{\sigma}\int \underline{F}_{\beta\alpha}(\underline{q}_\beta,\underline{q}_\alpha)\ \underline{c}_{\beta|\alpha}^{(2)}(\underline{q}_\beta\mid\underline{q}_\alpha;\underline{t})\ \underline{d}^6\underline{q}_\beta \qquad (27)$$

where $\underline{F}_\alpha^{(e)}(\underline{q}_\alpha;\underline{t})$ is the force exerted on the molecule of species α in configuration $\underline{q}_\alpha$ by the external field; and where $\underline{F}_{\beta\alpha}(\underline{q}_\beta,\underline{q}_\alpha)$ is the force exerted on the molecule of species α at $\underline{q}_\alpha$ by a molecule of species β at $\underline{q}_\beta$. Analogously, the total torque acting on a molecule of species α in configuration $\underline{q}_\alpha$, averaged over all configurations of the $N - 1$ remaining molecules of the system, is

$$\mathbf{T}_\alpha^{(1)}(\mathbf{q}_\alpha; t) = \frac{\int \mathbf{T}_{\alpha i}\ \delta(\mathbf{q}_{\alpha i} - \mathbf{q}_\alpha)\ f^{(N)}(\mathbf{q}; t)\ d\mathbf{q}}{\int \delta(\mathbf{q}_{\alpha i} - \mathbf{q}_\alpha)\ f^{(N)}(\mathbf{q}; t)\ d\mathbf{q}} \ ;$$

and for pair-additive molecular interactions this becomes

$$\mathbf{T}_\alpha^{(1)}(\mathbf{q}_\alpha; t) = \mathbf{T}_\alpha^{(e)}(\mathbf{q}_\alpha; t) + \sum_{\beta=1}^{\sigma} \int \mathbf{T}_{\beta\alpha}(\mathbf{q}_\beta, \mathbf{q}_\alpha)\ c_{\beta|\alpha}^{(2)}(\mathbf{q}_\beta \mid \mathbf{q}_\alpha; t)\ d^6\mathbf{q}_\beta \, . \tag{28}$$

For the total force acting at the center of mass of a molecule of species α at position $\mathbf{r}_\alpha$, averaged over all orientations of this molecule of species α as well as over all configurations of the $N - 1$ remaining molecules of the system, we write

$$\mathbf{F}_\alpha^{(1)}(\mathbf{r}_\alpha; t) = \frac{\int \mathbf{F}_{\alpha i}\ \delta(\mathbf{r}_{\alpha i} - \mathbf{r}_\alpha)\ f^{(N)}(\mathbf{q}; t)\ d\mathbf{q}}{\int \delta(\mathbf{r}_{\alpha i} - \mathbf{r}_\alpha)\ f^{(N)}(\mathbf{q}; t)\ d\mathbf{q}}$$

$$= \int \mathbf{F}_\alpha^{(1)}(\mathbf{q}_\alpha; t)\ f_\alpha^{(1)}(\mathbf{R}_\alpha \mid \mathbf{r}_\alpha; t)\ d^3\mathbf{R}_\alpha \, .$$

With pair-additive molecular interactions this becomes

$$\underset{\sim}{F}_\alpha^{(1)}(\underset{\sim}{r}_\alpha;\, t) = \int \underset{\sim}{F}_\alpha^{(e)}(\underset{\sim}{q}_\alpha;\, t)\; f_\alpha^{(1)}(\underset{\sim}{R}_\alpha \,|\, \underset{\sim}{r}_\alpha;\, t)\; d^3\underset{\sim}{R}_\alpha \; + \tag{29}$$

$$+ \sum_{\beta=1}^{\sigma} \iint \underset{\sim}{F}_{\beta\alpha}(\underset{\sim}{q}_\beta, \underset{\sim}{q}_\alpha)\; C_{\alpha\beta|\alpha}^{(2)}(\underset{\sim}{R}_\alpha; \underset{\sim}{q}_\beta \,|\, \underset{\sim}{r}_\alpha;\, t)\; d^3\underset{\sim}{R}_\alpha\; d^6\underset{\sim}{q}_\beta \, .$$

The total force acting at the center of mass of a molecule of species α in configuration $\underset{\sim}{q}_\alpha$, averaged over all configurations of the $N - 2$ molecules of the system other than the molecule of species α at $\underset{\sim}{q}_\alpha$ and a molecule of species β at $\underset{\sim}{q}_\beta$ is

$$\underset{\sim}{F}_{\alpha\beta,1}^{(2)}(\underset{\sim}{q}_\alpha; \underset{\sim}{q}_\beta;\, t) = \underset{\sim}{F}_{\beta\alpha,2}^{(2)}(\underset{\sim}{q}_\beta; \underset{\sim}{q}_\alpha;\, t) =$$

$$= \frac{\int \underset{\sim}{F}_{\alpha i}\; \delta(\underset{\sim}{q}_{\alpha i} - \underset{\sim}{q}_\alpha)\; \delta(\underset{\sim}{q}_{\beta j} - \underset{\sim}{q}_\beta)\; f^{(N)}(\underset{\sim}{q};\, t)\; d\underset{\sim}{q}}{\int \delta(\underset{\sim}{q}_{\alpha i} - \underset{\sim}{q}_\alpha)\; \delta(\underset{\sim}{q}_{\beta j} - \underset{\sim}{q}_\beta)\; f^{(N)}(\underset{\sim}{q};\, t)\; d\underset{\sim}{q}} \; ;$$

and for pair-additive molecular interactions this becomes

$$\underset{\sim}{F}_{\alpha\beta,1}^{(2)}(\underset{\sim}{q}_\alpha; \underset{\sim}{q}_\beta;\, t) = \underset{\sim}{F}_\alpha^{(e)}(\underset{\sim}{q}_\alpha;\, t) + \underset{\sim}{F}_{\beta\alpha}(\underset{\sim}{q}_\beta, \underset{\sim}{q}_\alpha) \; +$$

$$+ \sum_{\gamma=1}^{\sigma} \int \underset{\sim}{F}_{\gamma\alpha}(\underset{\sim}{q}_\gamma, \underset{\sim}{q}_\alpha)\; C_{\gamma|\alpha\beta}^{(3)}(\underset{\sim}{q}_\gamma \,|\, \underset{\sim}{q}_\alpha;\, \underset{\sim}{q}_\beta;\, t)\; d^6\underset{\sim}{q}_\gamma \, .$$

Similar expressions can be given for $\underset{\sim}{\underline{T}}_{\alpha\beta,1}^{(2)}(\underset{\sim}{\underline{q}}_\alpha; \underset{\sim}{\underline{q}}_\beta; \underline{t})$ and $\underset{\sim}{\underline{T}}_{\alpha\beta,2}^{(2)}(\underset{\sim}{\underline{q}}_\alpha; \underset{\sim}{\underline{q}}_\beta; \underline{t})$. Lastly, the total force acting on a molecule of species α at position $\underset{\sim}{\underline{r}}_\alpha$ there being a molecule of species β at position $\underset{\sim}{\underline{r}}_\beta$, averaged over all orientations of these molecules as well as over all configurations of the $\underline{N} - 2$ remaining molecules of the system, is

$$\underset{\sim}{\underline{F}}_{\alpha\beta,1}^{(2)}(\underset{\sim}{\underline{r}}_\alpha;\underset{\sim}{\underline{r}}_\beta;\underline{t}) = \underset{\sim}{\underline{F}}_{\beta\alpha,2}^{(2)}(\underset{\sim}{\underline{r}}_\beta;\underset{\sim}{\underline{r}}_\alpha;\underline{t}) =$$

$$= \frac{\int \underset{\sim}{\underline{F}}_{\alpha\underline{i}}\ \delta(\underset{\sim}{\underline{r}}_{\alpha\underline{i}} - \underset{\sim}{\underline{r}}_\alpha)\ \delta(\underset{\sim}{\underline{r}}_{\beta\underline{j}} - \underset{\sim}{\underline{r}}_\beta)\ \underline{f}^{(\underline{N})}(\underset{\sim}{\underline{q}};\underline{t})\ \underline{d}\underset{\sim}{\underline{q}}}{\int \delta(\underset{\sim}{\underline{r}}_{\alpha\underline{i}} - \underset{\sim}{\underline{r}}_\alpha)\ \delta(\underset{\sim}{\underline{r}}_{\beta\underline{j}} - \underset{\sim}{\underline{r}}_\beta)\ \underline{f}^{(\underline{N})}(\underset{\sim}{\underline{q}};\underline{t})\ \underline{d}\underset{\sim}{\underline{q}}}$$

$$= \iint \underset{\sim}{\underline{F}}_{\alpha\beta,1}^{(2)}(\underset{\sim}{\underline{q}}_\alpha;\underset{\sim}{\underline{q}}_\beta;\underline{t})\ \underline{f}_{\alpha\beta|\alpha\beta}^{(2)}(\underset{\sim}{\underline{R}}_\alpha;\underset{\sim}{\underline{R}}_\beta|\underset{\sim}{\underline{r}}_\alpha;\underset{\sim}{\underline{r}}_\beta;\underline{t})\ \underline{d}^3\underset{\sim}{\underline{R}}_\alpha\ \underline{d}^3\underset{\sim}{\underline{R}}_\beta.$$

For pair-additive molecular interactions this becomes

$$\underset{\sim}{\underline{F}}_{\alpha\beta,1}^{(2)}(\underset{\sim}{\underline{r}}_\alpha;\underset{\sim}{\underline{r}}_\beta;\underline{t}) = \int \underset{\sim}{\underline{F}}_\alpha^{(e)}(\underset{\sim}{\underline{q}}_\alpha;\underline{t})\ \underline{f}_{\alpha|\alpha\beta}^{(2)}(\underset{\sim}{\underline{R}}_\alpha|\underset{\sim}{\underline{r}}_\alpha;\underset{\sim}{\underline{r}}_\beta;\underline{t})\ \underline{d}^3\underset{\sim}{\underline{R}}_\alpha +$$

$$+ \iint \underset{\sim}{\underline{F}}_{\beta\alpha}(\underset{\sim}{\underline{q}}_\beta,\underset{\sim}{\underline{q}}_\alpha)\ \underline{f}_{\alpha\beta|\alpha\beta}^{(2)}(\underset{\sim}{\underline{R}}_\alpha;\underset{\sim}{\underline{R}}_\beta|\underset{\sim}{\underline{r}}_\alpha;\underset{\sim}{\underline{r}}_\beta;\underline{t})\ \underline{d}^3\underset{\sim}{\underline{R}}_\alpha\ \underline{d}^3\underset{\sim}{\underline{R}}_\beta +$$

$$+ \sum_{\gamma=1}^{\sigma} \iint \underset{\sim}{\underline{F}}_{\gamma\alpha}(\underset{\sim}{\underline{q}}_\gamma,\underset{\sim}{\underline{q}}_\alpha)\ \underline{C}_{\alpha\beta\gamma|\alpha\beta}^{(3)}(\underset{\sim}{\underline{R}}_\alpha;\underset{\sim}{\underline{R}}_\beta;\underset{\sim}{\underline{q}}_\gamma|\underset{\sim}{\underline{r}}_\alpha;\underset{\sim}{\underline{r}}_\beta;\underline{t})\ \underline{d}^3\underset{\sim}{\underline{R}}_\alpha\ \underline{d}^3\underset{\sim}{\underline{R}}_\beta\ \underline{d}^6\underset{\sim}{\underline{q}}_\gamma.$$

We shall now consider the task of indicating how theoretical distribution functions can be calculated.

Canonical Ensemble Distribution Functions

The canonical ensemble specific distribution function in the $12\underline{N}$-dimensional phase space of the $\underline{N}$ molecule system is given by[21, 22]

$$\underline{f}^{(\underline{N},0)}(\underset{\sim}{\underline{q}}, \underset{\sim}{\underline{p}}) = \left[\prod_{\alpha=1}^{\sigma} (\underline{N}_\alpha!\ \underline{h}^{6\underline{N}_\alpha})\ \underline{Q}_{\underline{N}}\right]^{-1} \exp\left[-\frac{1}{\underline{kT}}\ \mathcal{H}(\underline{q},\underline{p})\right]$$

where the partition function

$$\underline{Q}_{\underline{N}} = \left[\prod_{\alpha=1}^{\sigma} (\underline{N}_\alpha!\ \underline{h}^{6\underline{N}_\alpha})\right]^{-1} \int\!\!\int \exp\left[-\frac{1}{\underline{kT}}\ \mathcal{H}(\underline{q},\underline{p})\right] \underline{d}\,\underline{q}\ \ \underline{d}\,\underline{p}$$

is chosen so that $\underline{f}^{(\underline{N},0)}(\underline{q}, \underline{p})$ is normalized to unity. Here $\mathcal{H}(\underline{q}, \underline{p}) = \underline{T}(\underline{q}, \underline{p}) + \underline{U}(\underline{q})$ is the classical Hamiltonian function equal to the sum of the kinetic energy of the system and the potential energy of the system. This distribution function is appropriate for a classical fluid system containing $\underline{N}$ molecules in a state of macroscopic equilibrium in the volume $\underline{V}$ at temperature $\underline{T}$. Henceforth, the superscript (, 0) on a distribution function, correlation function, or average value of a dynamical variable will indicate that it is a canonical ensemble (equilibrium) distribution function, correlation function, or average value.

Integration over all the momenta gives the canonical ensemble specific distribution function in the 6N-dimensional configuration space of the N molecule system:

$$f^{(N,0)}(q) = \int f^{(N,0)}(q, p)\, dp = Z_N^{-1} \exp\left[-\frac{1}{kT}\, U(q)\right]$$

where Z_N is the configurational integral for this system as given by

$$Z_N = \int \exp\left[-\frac{1}{kT}\, U(q)\right] dq .$$

The canonical ensemble generic singlet distribution function in α-configuration space is

$$c_\alpha^{(1,0)}(q_\alpha) = N_\alpha\, f_{\alpha i}^{(1,0)}(q_\alpha) =$$

$$= \frac{N_\alpha}{Z_N} \int \delta(q_{\alpha i} - q_\alpha)\, \exp\left[-\frac{1}{kT}\, U(q)\right] dq .$$

For a fluid in the absence of external force fields, $\underline{C}_\alpha^{(1,0)}(\underline{q}_\alpha)$ is a constant independent of both $\underline{r}_\alpha$ and $\underline{R}_\alpha$. Normalization gives $\iint \underline{C}_\alpha^{(1,0)}(\underline{q}_\alpha)\, d^3\underline{r}_\alpha\, d^3\underline{R}_\alpha = 8\pi^2 \underline{V}\, \underline{C}_\alpha^{(1,0)}(\underline{q}_\alpha) = \underline{N}_\alpha$ so that $\underline{C}_\alpha^{(1,0)}(\underline{q}_\alpha) = \underline{N}_\alpha / 8\pi^2 \underline{V} = \underline{C}_\alpha / 8\pi^2$ where $\underline{C}_\alpha$ is the (constant) macroscopic number density of the α molecules. We have neglected the dependence of $\underline{C}_\alpha^{(1,0)}(\underline{q}_\alpha)$ on position within a region of negligible volume near the walls of the container of volume $\underline{V}$. Alternately, we may consider the volume $\underline{V}$ and the matter contained therein to be part of an infinite system. Other implications for a fluid in the absence of external force fields are that $\underline{C}_\alpha^{(1,0)}(\underline{r}_\alpha) = \underline{C}_\alpha$ and $\underline{C}_\alpha^{(1,0)}(\underline{R}_\alpha) = \underline{N}_\alpha / 8\pi^2$. Each of these canonical ensemble generic α-singlet distribution functions is normalized to $\underline{N}_\alpha$.

The <u>canonical ensemble generic pair distribution</u> function in $\alpha\beta$-pair configuration space is

$$\underline{C}_{\alpha\beta}^{(2,0)}(\underline{q}_\alpha;\, \underline{q}_\beta) = \underline{N}_\alpha (\underline{N}_\beta - \delta_{\alpha\beta})\, \underline{f}_{\alpha \underline{i} \beta \underline{j}}^{(2,0)}(\underline{q}_\alpha;\, \underline{q}_\beta) =$$

$$= \frac{\underline{N}_\alpha(\underline{N}_\beta - \delta_{\alpha\beta})}{\underline{Z}_{\underline{N}}} \int \delta(\underline{q}_{\alpha \underline{i}} - \underline{q}_\alpha)\, \delta(\underline{q}_{\beta \underline{j}} - \underline{q}_\beta)\, \exp\left[-\frac{1}{\underline{kT}}\, \underline{U}(\underline{q})\right] d\underline{q}\,.$$

As in equation (15), the <u>canonical ensemble</u> $\alpha\beta$-<u>pair</u>

configuration correlation function $\underline{g}_{\alpha\beta}^{(2,0)}(\underset{\sim}{\underline{q}}_{\alpha};\ \underset{\sim}{\underline{q}}_{\beta})$ is defined by

$$\underline{C}_{\alpha\beta}^{(2,0)}(\underset{\sim}{\underline{q}}_{\alpha};\underset{\sim}{\underline{q}}_{\beta}) = \underline{C}_{\alpha}^{(1,0)}(\underset{\sim}{\underline{q}}_{\alpha})\ \underline{C}_{\beta}^{(1,0)}(\underset{\sim}{\underline{q}}_{\beta})\ \underline{g}_{\alpha\beta}^{(2,0)}(\underset{\sim}{\underline{q}}_{\alpha};\underset{\sim}{\underline{q}}_{\beta}). \quad (30)$$

The boundary conditions on $\underline{g}_{\alpha\beta}^{(2,0)}(\underset{\sim}{\underline{q}}_{\alpha};\underset{\sim}{\underline{q}}_{\beta})$ are of course: (1) no correlation at infinite separation of the molecules in a fluid — see equation (16); and (2) no interpenetration of convex rigid molecules — see equations (17) and (18). The canonical ensemble generic pair distribution function in $\alpha\beta$-pair position space is obtained by the integration

$$\underline{C}_{\alpha\beta}^{(2,0)}(\underset{\sim}{\underline{r}}_{\alpha};\underset{\sim}{\underline{r}}_{\beta}) = \iint \underline{C}_{\alpha\beta}^{(2,0)}(\underset{\sim}{\underline{q}}_{\alpha};\underset{\sim}{\underline{q}}_{\beta})\ \underline{d}^3\underset{\sim}{\underline{R}}_{\alpha}\ \underline{d}^3\underset{\sim}{\underline{R}}_{\beta};$$

and, as in equation (19), the canonical ensemble $\alpha\beta$-pair position correlation function $\underline{g}_{\alpha\beta}^{(2,0)}(\underset{\sim}{\underline{r}}_{\alpha};\underset{\sim}{\underline{r}}_{\beta})$ is defined by

$$\underline{C}_{\alpha\beta}^{(2,0)}(\underset{\sim}{\underline{r}}_{\alpha};\underset{\sim}{\underline{r}}_{\beta}) = \underline{C}_{\alpha}^{(1,0)}(\underset{\sim}{\underline{r}}_{\alpha})\ \underline{C}_{\beta}^{(1,0)}(\underset{\sim}{\underline{r}}_{\beta})\ \underline{g}_{\alpha\beta}^{(2,0)}(\underset{\sim}{\underline{r}}_{\alpha};\underset{\sim}{\underline{r}}_{\beta}). \quad (31)$$

The boundary conditions on $\underline{g}_{\alpha\beta}^{(2,0)}(\underset{\sim}{\underline{r}}_{\alpha};\underset{\sim}{\underline{r}}_{\beta})$ are the same as those on $\underline{g}_{\alpha\beta}^{(2,0)}(\underset{\sim}{\underline{q}}_{\alpha};\underset{\sim}{\underline{q}}_{\beta})$. For a fluid in the absence of external force fields $\underline{c}_{\alpha\beta}^{(2,0)}(\underline{q}_\alpha;\underline{q}_\beta)$ and $\underline{g}_{\alpha\beta}^{(2,0)}(\underline{q}_\alpha;\underline{q}_\beta)$ become independent of the absolute position (boundary effects being neglected). They are functions of $\underline{R}_\alpha$, $\underline{R}_\beta$, and $\underline{r}_{\beta\alpha} = \underline{r}_\beta - \underline{r}_\alpha$ only. Similarly, $\underline{c}_{\alpha\beta}^{(2,0)}(\underline{r}_\alpha;\underline{r}_\beta)$ and $\underline{g}_{\alpha\beta}^{(2,0)}(\underline{r}_\alpha;\underline{r}_\beta)$ are functions of $\underline{r}_{\beta\alpha}$ only. Clearly, in this case

$$\underline{g}_{\alpha\beta}^{(2,0)}(\underline{r}_\alpha;\underline{r}_\beta) = \frac{1}{64\,\pi^4} \int \underline{g}_{\alpha\beta}^{(2,0)}(\underline{q}_\alpha;\underline{q}_\beta)\, \underline{d}^3\underline{R}_\alpha\, \underline{d}^3\underline{R}_\beta . \quad (32)$$

Of course each of these canonical ensemble generic $\alpha\beta$-pair distribution functions is normalized to $\underline{N}_\alpha(\underline{N}_\beta - \delta_{\alpha\beta})$.

The <u>potential of mean generalized forces</u> $\underline{W}_{\alpha\beta}^{(2,0)}(\underline{q}_\alpha;\underline{q}_\beta)$ associated with the generalized coordinates of a molecule of species α and a molecule of species β, averaged over all configurations of the remaining $\underline{N} - 2$ molecules of the system, is defined by[23]

$$\underline{g}_{\alpha\beta}^{(2,0)}(\underline{q}_\alpha;\underline{q}_\beta) = \left(\exp\left[-\frac{1}{\underline{kT}}\, \underline{W}_{\alpha\beta}^{(2,0)}(\underline{q}_\alpha;\underline{q}_\beta)\right]\right)\left(1 - \frac{\delta_{\alpha\beta}}{\underline{N}_\alpha}\right) . \quad (33)$$

Usually, $\delta_{\alpha\beta}/\underline{N}_\alpha$ will be neglected relative to unity. Upon using rectangular Cartesian coordinates (or spherical polar coordinates) and Eulerian angles for the generalized coordinates of the molecules of species α and β; and upon differentiating equation (33), we obtain the equations:

$$\underline{\underline{F}}_{\alpha\beta,1}^{(2,0)}(\underline{\underline{q}}_\alpha;\underline{\underline{q}}_\beta) - \underline{\underline{F}}_{\alpha}^{(1,0)}(\underline{\underline{q}}_\alpha) = -\nabla_{\underline{\underline{r}}_\alpha} \underline{W}_{\alpha\beta}^{(2,0)}(\underline{\underline{q}}_\alpha;\underline{\underline{q}}_\beta) \qquad (34)$$

$$\underline{\underline{F}}_{\alpha\beta,2}^{(2,0)}(\underline{\underline{q}}_\alpha;\underline{\underline{q}}_\beta) - \underline{\underline{F}}_{\beta}^{(1,0)}(\underline{\underline{q}}_\beta) = -\nabla_{\underline{\underline{r}}_\beta} \underline{W}_{\alpha\beta}^{(2,0)}(\underline{\underline{q}}_\alpha;\underline{\underline{q}}_\beta)$$

$$\underline{\underline{T}}_{\alpha\beta,1}^{(2,0)}(\underline{\underline{q}}_\alpha;\underline{\underline{q}}_\beta) - \underline{\underline{T}}_{\alpha}^{(1,0)}(\underline{\underline{q}}_\alpha) = -\nabla_{\underline{\underline{R}}_\alpha} \underline{W}_{\alpha\beta}^{(2,0)}(\underline{\underline{q}}_\alpha;\underline{\underline{q}}_\beta)$$

$$\underline{\underline{T}}_{\alpha\beta,2}^{(2,0)}(\underline{\underline{q}}_\alpha;\underline{\underline{q}}_\beta) - \underline{\underline{T}}_{\beta}^{(1,0)}(\underline{\underline{q}}_\beta) = -\nabla_{\underline{\underline{R}}_\beta} \underline{W}_{\alpha\beta}^{(2,0)}(\underline{\underline{q}}_\alpha;\underline{\underline{q}}_\beta)$$

where

$$\underline{\underline{F}}_{\alpha\beta,1}^{(2,0)}(\underline{\underline{q}}_\alpha;\underline{\underline{q}}_\beta) = \underline{\underline{F}}_{\beta\alpha,2}^{(2,0)}(\underline{\underline{q}}_\beta;\underline{\underline{q}}_\alpha) = \underline{kT}\nabla_{\underline{\underline{r}}_\alpha} \ln \underline{C}_{\alpha\beta}^{(2,0)}(\underline{\underline{q}}_\alpha;\underline{\underline{q}}_\beta)$$

$$\underline{\underline{T}}_{\alpha\beta,1}^{(2,0)}(\underline{\underline{q}}_\alpha;\underline{\underline{q}}_\beta) = \underline{\underline{T}}_{\beta\alpha,2}^{(2,0)}(\underline{\underline{q}}_\beta;\underline{\underline{q}}_\alpha) = \underline{kT}\nabla_{\underline{\underline{R}}_\alpha} \ln \underline{C}_{\alpha\beta}^{(2,0)}(\underline{\underline{q}}_\alpha;\underline{\underline{q}}_\beta)$$

$$\underline{\underline{F}}_{\alpha}^{(1,0)}(\underline{\underline{q}}_\alpha) = \underline{kT}\nabla_{\underline{\underline{r}}_\alpha} \ln \underline{C}_{\alpha}^{(1,0)}(\underline{\underline{q}}_\alpha)$$

$$(35)$$

$$\underline{\underline{T}}_{\alpha}^{(1,0)}(\underline{\underline{q}}_\alpha) = \underline{kT}\nabla_{\underline{\underline{R}}_\alpha} \ln \underline{C}_{\alpha}^{(1,0)}(\underline{\underline{q}}_\alpha).$$

For a fluid in the absence of external force fields $\underset{\approx}{F}_{\alpha}^{(1,0)}(\underset{\approx}{q}_{\alpha}) = 0$ and $\underset{\approx}{T}_{\alpha}^{(1,0)}(\underset{\approx}{q}_{\alpha}) = 0$; and since $\underline{W}_{\alpha\beta}^{(2,0)}(\underset{\approx}{q}_{\alpha};\underset{\approx}{q}_{\beta})$ is a function of $\underset{\approx}{R}_{\alpha}$, $\underset{\approx}{R}_{\beta}$, and $\underset{\approx}{r}_{\beta\alpha}$ only, introducing relative coordinates in position space gives

$$\underset{\approx}{F}_{\alpha\beta,1}^{(2,0)}(\underset{\approx}{q}_{\alpha};\underset{\approx}{q}_{\beta}) = -\underset{\approx}{F}_{\alpha\beta,2}^{(2,0)}(\underset{\approx}{q}_{\alpha};\underset{\approx}{q}_{\beta}) = \tag{36}$$

$$= \nabla_{\underset{\approx}{r}_{\beta\alpha}} \underline{W}_{\alpha\beta}^{(2,0)}(\underset{\approx}{q}_{\alpha};\underset{\approx}{q}_{\beta}).$$

Clearly, $\underline{W}_{\alpha\beta}^{(2,0)}(\underset{\approx}{r}_{\alpha};\underset{\approx}{r}_{\beta})$ can be defined in the usual way, and equations analogous to equations (33), (34), (35) and (36) can readily be obtained.

In order to obtain useful approximate expressions for $\underline{g}_{\alpha\beta}^{(2,0)}(\underset{\approx}{q}_{\alpha};\underset{\approx}{q}_{\beta})$ and $\underline{W}_{\alpha\beta}^{(2,0)}(\underset{\approx}{q}_{\alpha};\underset{\approx}{q}_{\beta})$, we introduce the <u>Kirkwood integral equation</u>[24]

$$\underline{W}_{\alpha\beta}^{(2,0)}(\underset{\approx}{q}_{\alpha};\underset{\approx}{q}_{\beta};\xi_{\beta}) = \xi_{\beta}\, \underline{V}_{\alpha\beta}(\underset{\approx}{q}_{\alpha},\underset{\approx}{q}_{\beta}) + \tag{37}$$

$$+ \sum_{\gamma=1}^{\sigma} \int_{0}^{\xi_{\beta}}\!\!\int \underline{V}_{\beta\gamma}(\underset{\approx}{q}_{\beta},\underset{\approx}{q}_{\gamma})\; \underline{c}_{\gamma}^{(1,0)}(\underset{\approx}{q}_{\gamma}) \left[\frac{\underline{g}_{\alpha\beta\gamma}^{(3,0)}(\underset{\approx}{q}_{\alpha};\underset{\approx}{q}_{\beta};\underset{\approx}{q}_{\gamma};\,\xi_{\beta})}{\underline{g}_{\alpha\beta}^{(2,0)}(\underset{\approx}{q}_{\alpha};\underset{\approx}{q}_{\beta};\,\xi_{\beta})} - \right.$$

$$\left. - \underline{g}_{\beta\gamma}^{(2,0)}(\underset{\approx}{q}_{\beta};\underset{\approx}{q}_{\gamma};\,\xi_{\beta})\right] \underline{d}\underset{\approx}{q}_{\gamma}\, \underline{d}\xi_{\beta} .$$

This is valid for a fluid at equilibrium in the absence of external force fields and with the neglect of terms of order $1/\underline{N}_\alpha$, where $\alpha = 1, \ldots, \sigma$, relative to unity. Here ξ_β, assigned to the molecule of species β at $\underline{q}_\beta$, is a coupling or charging parameter which ranges from 0, which represents the "uncharged state" of this molecule, to 1, which represents the "fully charged state" with full coupling of the intermolecular forces. Upon neglecting terms of order $\underline{C}_\gamma = \underline{N}_\gamma / \underline{V}$, where $\gamma = 1, \ldots, \sigma$, relative to unity; equation (37) gives an expression for the potential of mean force which is valid at "infinite dilution" or zero density

$$\underline{W}_{\alpha\beta}^{(2,0)}(\underline{q}_\alpha;\underline{q}_\beta) = \underline{V}_{\alpha\beta}(\underline{q}_\alpha,\underline{q}_\beta). \tag{38}$$

In this case,

$$\underline{g}_{\alpha\beta}^{(2,0)}(\underline{q}_\alpha;\underline{q}_\beta) = \exp\left[-\frac{1}{\underline{kT}}\ \underline{V}_{\alpha\beta}(\underline{q}_\alpha,\underline{q}_\beta)\right]. \tag{39}$$

The proper use of density expansions would, of course, result in better approximate expressions for $\underline{W}_{\alpha\beta}^{(2,0)}(\underline{q}_\alpha;\underline{q}_\beta)$ and $\underline{g}_{\alpha\beta}^{(2,0)}(\underline{q}_\alpha;\underline{q}_\beta)$.

Non-equilibrium Distribution Functions

An external force (for example, the force due to a homogeneous electric field $\underline{\underline{E}}$) sets up irreversible, transport processes within the system. Consequently the canonical ensemble distribution functions, which apply only to systems at equilibrium, are no longer valid. We must examine the behavior of the system when it is not at equilibrium, i.e., we must obtain expressions for non-equilibrium distribution functions.

We shall assume that the various non-equilibrium distribution functions consist of equilibrium terms of order zero in the external force field (that is, of canonical ensemble distribution functions applicable to the system when not subjected to the external force field) plus first order perturbations proportional to the external force field. Accordingly, we shall express macroscopic observables and average values of dynamical variables as sums of equilibrium terms and perturbation terms. In the practical cases which we shall consider, the external force fields will be so small that quadratic and higher order terms in them can be neglected.

The non-equilibrium generic singlet distribution function $\underline{C}_\alpha^{(1)}(\underline{\underline{q}}_\alpha; \underline{t})$ will, for example, be written as

$$\underline{c}_{\alpha}^{(1)}(\underline{q}_{\alpha};\,\underline{t}) \quad = \quad \underline{c}_{\alpha}^{(1,0)}(\underline{q}_{\alpha}) \quad + \quad \underline{c}_{\alpha}^{(1,1)}(\underline{q}_{\alpha};\,\underline{t})$$

where $\underline{c}_{\alpha}^{(1,0)}(\underline{q}_{\alpha})$ is an equilibrium term of order zero in the external field; and where $\underline{c}_{\alpha}^{(1,1)}(\underline{q}_{\alpha};\,\underline{t})$ is a perturbation term linear in the external field. The superscript (,0) now denotes an equilibrium term of order zero in the external field; and the superscript (,1) refers to a non-equilibrium perturbation term linear in the external field. Since both $\underline{c}_{\alpha}^{(1)}(\underline{q}_{\alpha};\,\underline{t})$ and $\underline{c}_{\alpha}^{(1,0)}(\underline{q}_{\alpha})$ are normalized to $\underline{N}_{\alpha}$, the perturbation term $\underline{c}_{\alpha}^{(1,1)}(\underline{q}_{\alpha};\,\underline{t})$ must be normalized to zero.

Similarly, the non-equilibrium generic pair distribution function $\underline{c}_{\alpha\beta}^{(2)}(\underline{q}_{\alpha};\underline{q}_{\beta};\,\underline{t})$ will be written as

$$\underline{c}_{\alpha\beta}^{(2)}(\underline{q}_{\alpha};\underline{q}_{\beta};\,\underline{t}) \quad = \quad \underline{c}_{\alpha\beta}^{(2,0)}(\underline{q}_{\alpha};\underline{q}_{\beta}) \;+\; \underline{c}_{\alpha\beta}^{(2,1)}(\underline{q}_{\alpha};\underline{q}_{\beta};\,\underline{t})$$

where $\underline{c}_{\alpha\beta}^{(2,0)}(\underline{q}_{\alpha};\underline{q}_{\beta})$ is the equilibrium term, and where $\underline{c}_{\alpha\beta}^{(2,1)}(\underline{q}_{\alpha};\underline{q}_{\beta};\,\underline{t})$ is a perturbation term which is normalized to zero.

The non-equilibrium $\alpha\beta$-pair configuration correlation function $\underline{G}_{\alpha\beta}^{(2)}(\underline{q}_{\alpha};\underline{q}_{\beta};\,\underline{t})$ is defined by equation (15). At equilibrium in the absence of external force fields, this correlation function is denoted by $\underline{g}_{\alpha\beta}^{(2,0)}(\underline{q}_{\alpha};\underline{q}_{\beta})$ and defined by equation (30). This suggests that we write

$$\underline{G}_{\alpha\beta}^{(2)}(\underline{q}_\alpha;\underline{q}_\beta;\underline{t}) = \underline{g}_{\alpha\beta}^{(2,0)}(\underline{q}_\alpha;\underline{q}_\beta) + \underline{G}_{\alpha\beta}^{(2,1)}(\underline{q}_\alpha;\underline{q}_\beta;\underline{t}) \tag{40}$$

where the equilibrium term $\underline{g}_{\alpha\beta}^{(2,0)}(\underline{q}_\alpha;\underline{q}_\beta)$ is of order zero in the external field and where $\underline{G}_{\alpha\beta}^{(2,1)}(\underline{q}_\alpha;\underline{q}_\beta;\underline{t})$ is a perturbation term linear in the external field. With the neglect of quadratic and higher order terms in the external field, we have

$$\underline{C}_{\alpha\beta}^{(2,1)}(\underline{q}_\alpha;\underline{q}_\beta;\underline{t}) = \underline{C}_{\alpha}^{(1,0)}(\underline{q}_\alpha)\,\underline{C}_{\beta}^{(1,0)}(\underline{q}_\beta)\;\underline{g}_{\alpha\beta}^{(2,1)}(\underline{q}_\alpha;\underline{q}_\beta;\underline{t}) \tag{41}$$

where

$$\underline{g}_{\alpha\beta}^{(2,1)}(\underline{q}_\alpha;\underline{q}_\beta;\underline{t}) = \underline{G}_{\alpha\beta}^{(2,1)}(\underline{q}_\alpha;\underline{q}_\beta;\underline{t}) + \tag{42}$$

$$+ \left[\frac{\underline{C}_{\alpha}^{(1,1)}(\underline{q}_\alpha;\underline{t})}{\underline{C}_{\alpha}^{(1,0)}(\underline{q}_\alpha)} + \frac{\underline{C}_{\beta}^{(1,1)}(\underline{q}_\beta;\underline{t})}{\underline{C}_{\beta}^{(1,0)}(\underline{q}_\beta)}\right] \underline{g}_{\alpha\beta}^{(2,0)}(\underline{q}_\alpha;\underline{q}_\beta) .$$

Frequently, it is convenient to use a non-equilibrium $\alpha\beta$-pair configuration correlation function $\underline{g}_{\alpha\beta}^{(2)}(\underline{q}_\alpha;\underline{q}_\beta;\underline{t})$ defined by

$$\underline{C}_{\alpha\beta}^{(2)}(\underline{q}_\alpha;\underline{q}_\beta;\underline{t}) = \underline{C}_\alpha^{(1,0)}(\underline{q}_\alpha)\, \underline{C}_\beta^{(1,0)}(\underline{q}_\beta)\, \underline{g}_{\alpha\beta}^{(2)}(\underline{q}_\alpha;\underline{q}_\beta;\underline{t}). \tag{43}$$

Clearly,

$$\underline{g}_{\alpha\beta}^{(2)}(\underline{q}_\alpha;\underline{q}_\beta;\underline{t}) = \underline{g}_{\alpha\beta}^{(2,0)}(\underline{q}_\alpha;\underline{q}_\beta) + \underline{g}_{\alpha\beta}^{(2,1)}(\underline{q}_\alpha;\underline{q}_\beta;\underline{t}) \tag{44}$$

so all deviations from equilibrium are "included" in the single term $\underline{g}_{\alpha\beta}^{(2,1)}(\underline{q}_\alpha;\underline{q}_\beta;\underline{t})$. As usual, the boundary conditions on $\underline{G}_{\alpha\beta}^{(2,1)}(\underline{q}_\alpha;\underline{q}_\beta;\underline{t})$ and $\underline{g}_{\alpha\beta}^{(2,1)}(\underline{q}_\alpha;\underline{q}_\beta;\underline{t})$ are: (1) no correlation at infinite separation of molecules in a fluid, whence

$$\lim_{\underline{r}_{\alpha\beta}\to +\infty} \underline{G}_{\alpha\beta}^{(2,1)}(\underline{q}_\alpha;\underline{q}_\beta;\underline{t}) = 0 \tag{45}$$

$$\lim_{\underline{r}_{\alpha\beta}\to +\infty} \underline{g}_{\alpha\beta}^{(2,1)}(\underline{q}_\alpha;\underline{q}_\beta;\underline{t}) = \tag{46}$$

$$= \left[\frac{\underline{C}_\alpha^{(1,1)}(\underline{q}_\alpha;\underline{t})}{\underline{C}_\alpha^{(1,0)}(\underline{q}_\alpha)} + \frac{\underline{C}_\beta^{(1,1)}(\underline{q}_\beta;\underline{t})}{\underline{C}_\beta^{(1,0)}(\underline{q}_\beta)}\right]\left[1 - \frac{\delta_{\alpha\beta}}{\underline{N}_\alpha}\right];$$

and (2) no interpenetration of convex rigid molecules, so

we have

$$\underline{G}_{\alpha\beta}^{(2,1)}(\underset{\sim}{\underline{q}}_{\alpha};\underset{\sim}{\underline{q}}_{\beta};\underline{t}) = 0 \qquad \text{if} \qquad \underline{r}_{\alpha\beta} < \underline{a}_{\alpha\beta}(\underset{\sim}{\underline{R}}_{\alpha},\underset{\sim}{\underline{R}}_{\beta};\underset{\sim}{\underline{1}}_{\underline{r}_{\beta\alpha}}) \tag{47}$$

$$\underline{g}_{\alpha\beta}^{(2,1)}(\underset{\sim}{\underline{q}}_{\alpha};\underset{\sim}{\underline{q}}_{\beta};\underline{t}) = 0 \qquad \text{if} \qquad \underline{r}_{\alpha\beta} < \underline{a}_{\alpha\beta}(\underset{\sim}{\underline{R}}_{\alpha},\underset{\sim}{\underline{R}}_{\beta};\underset{\sim}{\underline{1}}_{\underline{r}_{\beta\alpha}}), \tag{48}$$

as well as equation (26).

Analogously, we define and characterize the non-equilibrium $\alpha\beta$-pair position correlation functions $\underline{G}_{\alpha\beta}^{(2)}(\underset{\sim}{\underline{r}}_{\alpha};\underset{\sim}{\underline{r}}_{\beta};\underline{t})$ and $\underline{g}_{\alpha\beta}^{(2)}(\underset{\sim}{\underline{r}}_{\alpha};\underset{\sim}{\underline{r}}_{\beta};\underline{t})$, and the perturbations $\underline{G}_{\alpha\beta}^{(2,1)}(\underset{\sim}{\underline{r}}_{\alpha};\underset{\sim}{\underline{r}}_{\beta};\underline{t})$ and $\underline{g}_{\alpha\beta}^{(2,1)}(\underset{\sim}{\underline{r}}_{\alpha};\underset{\sim}{\underline{r}}_{\beta};\underline{t})$, by equations (19) and (31) and by equations similar to equations (40) through (48) as well as (26).

If both $\underline{C}_{\alpha}^{(1,1)}(\underset{\sim}{\underline{r}}_{\alpha};\underline{t})$ and $\underline{C}_{\beta}^{(1,1)}(\underset{\sim}{\underline{r}}_{\beta};\underline{t})$ are zero (as they are for the steady state of a fluid in a constant, homogeneous external field), then $\underline{g}_{\alpha\beta}^{(2,1)}(\underset{\sim}{\underline{r}}_{\alpha};\underset{\sim}{\underline{r}}_{\beta};\underline{t}) = \underline{G}_{\alpha\beta}^{(2,1)}(\underset{\sim}{\underline{r}}_{\alpha};\underset{\sim}{\underline{r}}_{\beta};\underline{t})$.

For a fluid at equilibrium in the absence of external force fields, $\underline{C}_{\alpha}^{(1,0)}(\underset{\sim}{\underline{q}}_{\alpha}) = (1/8\pi^2)\,\underline{C}_{\alpha}$ and $\underline{C}_{\alpha}^{(1,0)}(\underset{\sim}{\underline{r}}_{\alpha}) = \underline{C}_{\alpha}$. Consequently, we have

$$\underline{g}_{\alpha\beta}^{(2)}(\underset{\sim}{\underline{r}}_{\alpha};\underset{\sim}{\underline{r}}_{\beta};\underline{t}) = \frac{1}{64\pi^4}\iint \underline{g}_{\alpha\beta}^{(2)}(\underset{\sim}{\underline{q}}_{\alpha};\underset{\sim}{\underline{q}}_{\beta};\underline{t})\ \underline{d}^3\underset{\sim}{\underline{R}}_{\alpha}\ \underline{d}^3\underset{\sim}{\underline{R}}_{\beta}$$

and upon using equation (32) we obtain the result

$$\underline{g}_{\alpha\beta}^{(2,1)}(\underline{r}_\alpha;\underline{r}_\beta;\underline{t}) = \frac{1}{64\pi^4}\iint \underline{g}_{\alpha\beta}^{(2,1)}(\underline{q}_\alpha;\underline{q}_\beta;\underline{t})\ \underline{d}^3\underline{R}_\alpha\ \underline{d}^3\underline{R}_\beta. \quad (49)$$

Upon neglecting quadratic and higher order terms in the external field, the non-equilibrium conditional generic distribution function $\underline{c}_{\beta|\alpha}^{(2)}(\underline{q}_\beta|\underline{q}_\alpha;\underline{t})$ can be written as

$$\underline{c}_{\beta|\alpha}^{(2)}(\underline{q}_\beta|\underline{q}_\alpha;\underline{t}) = \underline{c}_{\beta|\alpha}^{(2,0)}(\underline{q}_\beta|\underline{q}_\alpha) + \underline{c}_{\beta|\alpha}^{(2,1)}(\underline{q}_\beta|\underline{q}_\alpha;\underline{t})$$

where

$$\underline{c}_{\beta|\alpha}^{(2,0)}(\underline{q}_\beta|\underline{q}_\alpha) = \underline{c}_\beta^{(1,0)}(\underline{q}_\beta)\ \underline{g}_{\alpha\beta}^{(2,0)}(\underline{q}_\alpha;\underline{q}_\beta)$$

is an equilibrium term of order zero in the external field, and where

$$\underline{c}_{\beta|\alpha}^{(2,1)}(\underline{q}_\beta|\underline{q}_\alpha;\underline{t}) = \underline{c}_\beta^{(1,0)}(\underline{q}_\beta)\ \underline{G}_{\alpha\beta}^{(2,1)}(\underline{q}_\alpha;\underline{q}_\beta;\underline{t}) + \\ + \underline{c}_\beta^{(1,1)}(\underline{q}_\beta;\underline{t})\ \underline{g}_{\alpha\beta}^{(2,0)}(\underline{q}_\alpha;\underline{q}_\beta)$$

$$= \underline{c}_{\beta}^{(1,0)}(\underset{\sim}{\underline{q}}_{\beta}) \left[\underline{g}_{\alpha\beta}^{(2,1)}(\underset{\sim}{\underline{q}}_{\alpha};\underset{\sim}{\underline{q}}_{\beta};\underline{t}) - \frac{\underline{c}_{\alpha}^{(1,1)}(\underset{\sim}{\underline{q}}_{\alpha};\underline{t})}{\underline{c}_{\alpha}^{(1,0)}(\underset{\sim}{\underline{q}}_{\alpha})} \underline{g}_{\alpha\beta}^{(2,0)}(\underset{\sim}{\underline{q}}_{\alpha};\underset{\sim}{\underline{q}}_{\beta}) \right]$$

is a perturbation term linear in the external field. For the non-equilibrium conditional generic distribution function $\underline{c}_{\alpha\beta|\alpha}^{(2)}(\underset{\sim}{\underline{R}}_{\alpha};\underset{\sim}{\underline{q}}_{\beta}|\underset{\sim}{\underline{r}}_{\alpha};\underline{t})$ we write

$$\underline{c}_{\alpha\beta|\alpha}^{(2)}(\underset{\sim}{\underline{R}}_{\alpha};\underset{\sim}{\underline{q}}_{\beta}|\underset{\sim}{\underline{r}}_{\alpha};\underline{t}) = \underline{c}_{\alpha\beta|\alpha}^{(2,0)}(\underset{\sim}{\underline{R}}_{\alpha};\underset{\sim}{\underline{q}}_{\beta}|\underset{\sim}{\underline{r}}_{\alpha}) + \underline{c}_{\alpha\beta|\alpha}^{(2,1)}(\underset{\sim}{\underline{R}}_{\alpha};\underset{\sim}{\underline{q}}_{\beta}|\underset{\sim}{\underline{r}}_{\alpha};\underline{t}) \tag{50}$$

where

$$\underline{c}_{\alpha\beta|\alpha}^{(2,0)}(\underset{\sim}{\underline{R}}_{\alpha};\underset{\sim}{\underline{q}}_{\beta}|\underset{\sim}{\underline{r}}_{\alpha}) = \frac{\underline{c}_{\alpha}^{(1,0)}(\underset{\sim}{\underline{q}}_{\alpha})\ \underline{c}_{\beta}^{(1,0)}(\underset{\sim}{\underline{q}}_{\beta})}{\underline{c}_{\alpha}^{(1,0)}(\underset{\sim}{\underline{r}}_{\alpha})} \underline{g}_{\alpha\beta}^{(2,0)}(\underset{\sim}{\underline{q}}_{\alpha};\underset{\sim}{\underline{q}}_{\beta}) \tag{51}$$

and

$$\underline{c}_{\alpha\beta|\alpha}^{(2,1)}(\underset{\sim}{\underline{R}}_{\alpha};\underset{\sim}{\underline{q}}_{\beta}|\underset{\sim}{\underline{r}}_{\alpha};\underline{t}) = \underline{c}_{\alpha}^{(1,0)}(\underset{\sim}{\underline{q}}_{\alpha})\ \underline{c}_{\beta}^{(1,0)}(\underset{\sim}{\underline{q}}_{\beta}) \left[\frac{\underline{g}_{\alpha\beta}^{(2)}(\underset{\sim}{\underline{q}}_{\alpha};\underset{\sim}{\underline{q}}_{\beta};\underline{t})}{\underline{c}_{\alpha}^{(1)}(\underset{\sim}{\underline{r}}_{\alpha};\underline{t})} - \frac{\underline{g}_{\alpha\beta}^{(2,0)}(\underset{\sim}{\underline{q}}_{\alpha};\underset{\sim}{\underline{q}}_{\beta})}{\underline{c}_{\alpha}^{(1,0)}(\underset{\sim}{\underline{r}}_{\alpha})} \right]. \tag{52}$$

If $\underset{\sim}{C}_\alpha^{(1,1)}(\underset{\sim}{r}_\alpha;\underline{t})$ happens to be zero (as it is for the steady
state of a fluid in a constant, homogeneous external field),
this last expression reduces to

$$\underset{\alpha\beta|\alpha}{C}^{(2,1)}(\underset{\sim}{R}_\alpha;\underset{\sim}{q}_\beta|\underset{\sim}{r}_\alpha;\underline{t}) = \frac{\underset{\sim}{C}_\alpha^{(1,0)}(\underset{\sim}{q}_\alpha)\ \underset{\sim}{C}_\beta^{(1,0)}(\underset{\sim}{q}_\beta)}{\underset{\sim}{C}_\alpha^{(1,0)}(\underset{\sim}{r}_\alpha)}\ \underset{\alpha\beta}{g}^{(2,1)}(\underset{\sim}{q}_\alpha;\underset{\sim}{q}_\beta;\underline{t}). \tag{53}$$

We shall now develop the theoretical apparatus needed in
order to obtain approximate expressions for non-equilibrium
distribution functions.

Equations of Continuity and Motion

On the hypothesis that no molecules are created or
destroyed, $\underline{f}^{(N)}(\underset{\sim}{q},\underset{\sim}{p};\underline{t})$ satisfies an equation of continuity
in the 12$\underline{N}$-dimensional phase space of the N molecule system.
This is the famous <u>Liouville equation</u>

$$\frac{d\underline{f}^{(N)}}{d\underline{t}} = \frac{\partial\underline{f}^{(N)}}{\partial\underline{t}} + \sum_{\alpha=1}^{\sigma}\sum_{i=1}^{N_\alpha}\sum_{k=1}^{6}\left(\frac{\partial\underline{f}^{(N)}}{\partial\underline{p}_{\alpha ik}}\dot{\underline{p}}_{\alpha ik} + \frac{\partial\underline{f}^{(N)}}{\partial\underline{q}_{\alpha ik}}\dot{\underline{q}}_{\alpha ik}\right) = 0 \tag{54}$$

which forms the basis of our non-equilibrium statistical
mechanical theory. Upon introducing rectangular Cartesian
coordinates, Eulerian angles, and the momenta conjugate to

them, we can write this equation in the form

$$\frac{\partial f^{(N)}}{\partial t} + \sum_{\alpha=1}^{\sigma} \sum_{i=1}^{N_\alpha} \left(\frac{\mathbf{m}_{\alpha i}}{m_\alpha} \cdot \nabla_{\mathbf{r}_{\alpha i}} + \mathbf{F}_{\alpha i} \cdot \nabla_{\mathbf{m}_{\alpha i}} + \boldsymbol{\omega}_{\alpha i} \cdot \nabla_{\mathbf{R}_{\alpha i}} + \mathbf{T}_{\alpha i} \cdot \nabla_{\mathbf{M}_{\alpha i}} \right) f^{(N)} = 0 \tag{55}$$

where

$$\nabla_{\mathbf{m}_{\alpha i}} = \mathbf{i} \frac{\partial}{\partial p_{x_{\alpha i}}} + \mathbf{j} \frac{\partial}{\partial p_{y_{\alpha i}}} + \mathbf{k} \frac{\partial}{\partial p_{z_{\alpha i}}} \tag{56}$$

is the gradient operator in the linear momentum space of the ith molecule of species α; and where

$$\nabla_{\mathbf{M}_{\alpha i}} = \mathbf{1}_{R_{\alpha i}} \left(\frac{\partial}{\partial p_{\Psi_{\alpha i}}} + \cos \Theta_{\alpha i} \frac{\partial}{\partial p_{\Phi_{\alpha i}}} \right) + \mathbf{1}_{\Theta_{\alpha i}} \left(- \sin \Theta_{\alpha i} \frac{\partial}{\partial p_{\Phi_{\alpha i}}} \right) + \mathbf{1}_{\Phi_{\alpha i}} \frac{\partial}{\partial p_{\Theta_{\alpha i}}} \tag{57}$$

is the gradient operator in the rotational momentum space

of the $\underline{i}^{th}$ molecule of species α.

We now let $\underline{G}(\underline{q},\underline{p})$ be any dynamical variable of the system which is not explicitly dependent on time. At time $\underline{t}$, the expectation value of $\underline{G}$ is given by equation (20). From this definition and the Liouville equation, the rate of change of the expectation value of $\underline{G}$ is found to be

$$\frac{\partial}{\partial \underline{t}} \left\langle \underline{G};\ \underline{f}^{(\underline{N})} \right\rangle = \left\langle \underline{G};\ \frac{\partial \underline{f}^{(\underline{N})}}{\partial \underline{t}} \right\rangle = \tag{58}$$

$$= - \sum_{\alpha=1}^{\sigma} \sum_{\underline{i}=1}^{\underline{N}_\alpha} \left\langle \underline{G}; \left[\frac{\underline{m}_{\alpha\underline{i}}}{\underline{m}_\alpha} \cdot \nabla_{\underline{r}_{\alpha\underline{i}}} + \underline{F}_{\alpha\underline{i}} \cdot \nabla_{\underline{m}_{\alpha\underline{i}}} + \right.\right.$$

$$\left.\left. + \ \omega_{\alpha\underline{i}} \cdot \nabla_{\underline{R}_{\alpha\underline{i}}} + \underline{T}_{\alpha\underline{i}} \cdot \nabla_{\underline{M}_{\alpha\underline{i}}} \right] \underline{f}^{(\underline{N})} \right\rangle .$$

By applying Green's theorem to the position space, to the orientation space, to the linear momentum space, and to the rotational momentum space of the $\underline{i}^{th}$ molecule of species α; and by neglecting surface integrals over the boundaries of these spaces since they have integrands containing $\underline{f}^{(\underline{N})}$, we can cast the previous equation of change into the form

$$\frac{\partial}{\partial \underline{t}} \left\langle \underline{G};\ \underline{f}^{(\underline{N})} \right\rangle = \sum_{\alpha=1}^{\sigma} \sum_{\underline{i}=1}^{\underline{N}_\alpha} \left\langle \left[\left(\frac{\underline{m}_{\alpha\underline{i}}}{\underline{m}_\alpha} \oplus \omega_{\alpha\underline{i}} \right) \cdot \left(\nabla_{\underline{r}_{\alpha\underline{i}}} \oplus \nabla_{\underline{R}_{\alpha\underline{i}}} \right) + \right.\right. \tag{59}$$

$$+ \left(\underline{\mathbf{F}}_{\alpha \underline{i}} \oplus \underline{\mathbf{T}}_{\alpha \underline{i}}\right) \cdot \left(\nabla_{\underline{\mathbf{m}}_{\alpha \underline{i}}} \oplus \nabla_{\underline{\mathbf{M}}_{\alpha \underline{i}}}\right) \Big] \underline{G};\ \underline{f}^{(\underline{N})} \Big\rangle .$$

This result is known as the <u>general equation of change</u>. From it we can construct the equation of change for the expectation value $\langle \underline{G};\ \underline{f}^{(\underline{N})} \rangle$ of any dynamical variable $\underline{G}$. We shall now utilize it in deriving the equations of continuity and motion in singlet space and in pair space.

To derive the <u>equation of continuity in</u> α-<u>configuration space</u>, we choose

$$\underline{G} = \sum_{\underline{i}=1}^{\underline{N}_\alpha} \delta(\underline{\mathbf{q}}_{\alpha \underline{i}} - \underline{\mathbf{q}}_\alpha) = \sum_{\underline{i}=1}^{\underline{N}_\alpha} \delta(\underline{\mathbf{r}}_{\alpha \underline{i}} - \underline{\mathbf{r}}_\alpha)\ \delta(\underline{\mathbf{R}}_{\alpha \underline{i}} - \underline{\mathbf{R}}_\alpha)$$

so that its expectation value $\langle \underline{G};\ \underline{f}^{(\underline{N})} \rangle$ is the singlet generic distribution function $\underline{C}_\alpha^{(1)}(\underline{\mathbf{q}}_\alpha;\underline{t})$. The equation of change then becomes the desired equation

$$\frac{\partial \underline{C}_\alpha^{(1)}(\underline{\mathbf{q}}_\alpha;\underline{t})}{\partial \underline{t}} = - \nabla_{\underline{\mathbf{r}}_\alpha} \cdot \left[\underline{C}_\alpha^{(1)}(\underline{\mathbf{q}}_\alpha;\underline{t})\ \underline{\mathbf{u}}_\alpha(\underline{\mathbf{q}}_\alpha;\underline{t})\right] - \nabla_{\underline{\mathbf{R}}_\alpha} \cdot \left[\underline{C}_\alpha^{(1)}(\underline{\mathbf{q}}_\alpha;\underline{t})\ \mathbf{\Omega}_\alpha(\underline{\mathbf{q}}_\alpha;\underline{t})\right] \tag{60}$$

which can be used in calculating $\underline{C}_\alpha^{(1)}(\underline{\mathbf{q}}_\alpha;\underline{t})$ if one can obtain usable expressions for $\underset{\sim}{u}_\alpha(\underset{\approx}{q}_\alpha;\underline{t})$ and $\underset{\sim}{\Omega}_\alpha(\underset{\approx}{q}_\alpha;\underline{t})$.

By choosing

$$\underline{G} = \sum_{\underline{i}=1}^{\underline{N}_\alpha} \delta(\underset{\approx}{r}_{\alpha\underline{i}} - \underset{\approx}{r}_\alpha)$$

so that $\langle \underline{G};\ \underline{f}^{(\underline{N})} \rangle$ is $\underline{C}_\alpha^{(1)}(\underset{\approx}{r}_\alpha;\underline{t})$, the equation of change becomes the equation of continuity in α-position space

$$\frac{\partial \underline{C}_\alpha^{(1)}(\underset{\approx}{r}_\alpha;\underline{t})}{\partial \underline{t}} = - \nabla_{\underset{\approx}{r}_\alpha} \cdot \left[\underline{C}_\alpha^{(1)}(\underset{\approx}{r}_\alpha;\underline{t})\ \underset{\sim}{u}_\alpha(\underset{\approx}{r}_\alpha;\underline{t}) \right]. \tag{61}$$

Upon multiplying this last equation by $\underline{m}_\alpha$ and summing over all species we obtain the total equation of continuity

$$\frac{\partial\ \rho(\underset{\approx}{r}_\alpha;\underline{t})}{\partial \underline{t}} = - \nabla_{\underset{\approx}{r}_\alpha} \cdot \left[\rho(\underset{\approx}{r}_\alpha;\underline{t})\ \underset{\sim}{u}(\underset{\approx}{r}_\alpha;\underline{t}) \right]. \tag{62}$$

To derive the equation of continuity in αβ-pair configuration space we choose

$$\underline{G} = \sum_{\underline{i}=1}^{\underline{N}_\alpha} \sum_{\substack{\underline{j}=1 \\ \alpha\underline{i} \neq \beta\underline{j}}}^{\underline{N}_\beta} \delta(\underset{\approx}{q}_{\alpha\underline{i}} - \underset{\approx}{q}_\alpha)\ \delta(\underset{\approx}{q}_{\beta\underline{j}} - \underset{\approx}{q}_\beta)$$

so that its expectation value $\langle G; f^{(N)} \rangle$ is the generic pair distribution function $C_{\alpha\beta}^{(2)}(q_\alpha;q_\beta;t)$. The equation of change then becomes

$$\frac{\partial C_{\alpha\beta}^{(2)}(q_\alpha;q_\beta;t)}{\partial t} = -\nabla_{r_\alpha} \cdot \left[C_{\alpha\beta}^{(2)}(q_\alpha;q_\beta;t)\ u_{\alpha\beta,1}^{(2)}(q_\alpha;q_\beta;t)\right] \tag{63}$$

$$- \nabla_{r_\beta} \cdot \left[C_{\alpha\beta}^{(2)}(q_\alpha;q_\beta;t)\ u_{\alpha\beta,2}^{(2)}(q_\alpha;q_\beta;t)\right]$$

$$- \nabla_{R_\alpha} \cdot \left[C_{\alpha\beta}^{(2)}(q_\alpha;q_\beta;t)\ \Omega_{\alpha\beta,1}^{(2)}(q_\alpha;q_\beta;t)\right]$$

$$- \nabla_{R_\beta} \cdot \left[C_{\alpha\beta}^{(2)}(q_\alpha;q_\beta;t)\ \Omega_{\alpha\beta,2}^{(2)}(q_\alpha;q_\beta;t)\right] .$$

This important equation of continuity can be used in calculating $C_{\alpha\beta}^{(2)}(q_\alpha;q_\beta;t)$ provided usable expressions can be obtained for the mean local velocities.

The equation of continuity in $\alpha\beta$-pair position space is

$$\frac{\partial C_{\alpha\beta}^{(2)}(r_\alpha;r_\beta;t)}{\partial t} = -\nabla_{r_\alpha} \cdot \left[C_{\alpha\beta}^{(2)}(r_\alpha;r_\beta;t)\ u_{\alpha\beta,1}^{(2)}(r_\alpha;r_\beta;t)\right] \tag{64}$$

$$- \nabla_{r_\beta} \cdot \left[C_{\alpha\beta}^{(2)}(r_\alpha;r_\beta;t)\ u_{\alpha\beta,2}^{(2)}(r_\alpha;r_\beta;t)\right].$$

This equation was solved by Onsager and Fuoss in their development of the theory of the electrical conductance of dilute solutions of simple ions.

The equation of motion in α-configuration space is obtained by writing the general equation of change with the dynamical variable

$$\underline{G} = \sum_{\underline{i}=1}^{\underline{N}_\alpha} (\underline{\underline{m}}_{\alpha\underline{i}} \oplus \underline{\underline{L}}_{\alpha\underline{i}}) \, \delta(\underline{\underline{q}}_{\alpha\underline{i}} - \underline{\underline{q}}_\alpha) \, .$$

With the temperature $\underline{T}(\underline{\underline{q}}_\alpha;\underline{t})$ defined in terms of the average kinetic energy of a molecule of species α by the relationship[25]

$$3\underline{k}\,\underline{T}(\underline{\underline{q}}_\alpha;\underline{t}) = \int \frac{\underline{\underline{m}}_\alpha \cdot \underline{\underline{m}}_\alpha}{2\,\underline{m}_\alpha} \, \underline{f}_\alpha^{(1)}(\underline{\underline{m}}_\alpha | \underline{\underline{q}}_\alpha;\underline{t}) \, \underline{d}^3\underline{\underline{m}}_\alpha \; -$$

$$- \frac{1}{2}\, \underline{m}_\alpha \, \underline{\underline{u}}_\alpha(\underline{\underline{q}}_\alpha;\underline{t}) \cdot \underline{\underline{u}}_\alpha(\underline{\underline{q}}_\alpha;\underline{t}) \; +$$

$$+ \frac{1}{2} \int \underline{\omega}_\alpha \cdot \underline{\underline{L}}_\alpha \, \underline{f}_\alpha^{(1)}(\underline{\underline{M}}_\alpha | \underline{\underline{q}}_\alpha;\underline{t}) \, \underline{d}^3\underline{\underline{M}}_\alpha \; -$$

$$- \frac{1}{2}\, \underline{\Omega}_\alpha(\underline{\underline{q}}_\alpha;\underline{t}) \cdot \underline{\underline{I}}_\alpha \cdot \underline{\Omega}_\alpha(\underline{\underline{q}}_\alpha;\underline{t}) \, ;$$

and with the substantial derivative given by

$$\frac{d}{d\underline{t}} = \frac{\partial}{\partial \underline{t}} + \left[\underline{u}_\alpha(\underline{q}_\alpha;\underline{t}) \oplus \Omega_\alpha(\underline{q}_\alpha;\underline{t})\right] \cdot \left[\nabla_{\underline{r}_\alpha} \oplus \nabla_{\underline{R}_\alpha}\right] ;$$

this equation of motion yields the two equations:

$$\frac{d}{d\underline{t}}\left[\underline{m}_\alpha \ \underline{u}_\alpha(\underline{q}_\alpha;\underline{t})\right] = -\nabla_{\underline{r}_\alpha}\left[\underline{kT} \ln \underline{c}_\alpha^{(1)}(\underline{q}_\alpha;\underline{t})\right] + \underline{F}_\alpha^{(1)}(\underline{q}_\alpha;\underline{t}) \tag{65}$$

$$\frac{d}{d\underline{t}}\left[\Lambda_\alpha(\underline{q}_\alpha;\underline{t})\right] = -\nabla_{\underline{R}_\alpha}\left[\underline{kT} \ln \underline{c}_\alpha^{(1)}(\underline{q}_\alpha;\underline{t})\right] + \underline{T}_\alpha^{(1)}(\underline{q}_\alpha;\underline{t}).$$

In a similar way, one can derive the <u>equation of motion in</u> α-<u>position space</u>

$$\frac{d}{d\underline{t}}\left[\underline{m}_\alpha \ \underline{u}_\alpha(\underline{r}_\alpha;\underline{t})\right] = -\nabla_{\underline{r}_\alpha}\left[\underline{kT} \ln \underline{c}_\alpha^{(1)}(\underline{r}_\alpha;\underline{t})\right] + \underline{F}_\alpha^{(1)}(\underline{r}_\alpha;\underline{t}). \tag{66}$$

To obtain the <u>equation of motion in</u> αβ-<u>pair configuration space</u> we choose

$$\underline{G} = \sum_{\substack{\underline{i}=1 \\ \alpha\underline{i} \neq \beta\underline{j}}}^{N_\alpha} \sum_{\underline{j}=1}^{N_\beta} \left[(\underline{m}_{\alpha\underline{i}} \oplus \underline{L}_{\alpha\underline{i}}) \oplus (\underline{m}_{\beta\underline{j}} \oplus \underline{L}_{\beta\underline{j}})\right] \delta(\underline{q}_{\alpha\underline{i}} - \underline{q}_\alpha)\, \delta(\underline{q}_{\beta\underline{j}} - \underline{q}_\beta)$$

Upon defining the temperature $\underline{T}(\underset{\sim}{\underline{q}}_\alpha;\underset{\sim}{\underline{q}}_\beta;\underline{t})$ in terms of the average kinetic energy in pair space of a molecule of species α and a molecule of species β; and upon introducing the substantial derivative in pair space; we obtain, after considerable manipulation, the following equations:

$$\frac{d}{d\,t}\left[\underline{m}_\alpha\, \underset{\sim}{\underline{u}}_{\alpha\beta,1}^{(2)}(\underset{\sim}{\underline{q}}_\alpha;\underset{\sim}{\underline{q}}_\beta;\underline{t})\right] = -\nabla_{\underset{\sim}{\underline{r}}_\alpha}\left[\underline{kT}\ln \underline{C}_{\alpha\beta}^{(2)}(\underset{\sim}{\underline{q}}_\alpha;\underset{\sim}{\underline{q}}_\beta;\underline{t})\right] + \underset{\sim}{\underline{F}}_{\alpha\beta,1}^{(2)}(\underset{\sim}{\underline{q}}_\alpha;\underset{\sim}{\underline{q}}_\beta;\underline{t}) \tag{67}$$

$$\frac{d}{d\,t}\left[\underline{m}_\beta\, \underset{\sim}{\underline{u}}_{\alpha\beta,2}^{(2)}(\underset{\sim}{\underline{q}}_\alpha;\underset{\sim}{\underline{q}}_\beta;\underline{t})\right] = -\nabla_{\underset{\sim}{\underline{r}}_\beta}\left[\underline{kT}\ln \underline{C}_{\alpha\beta}^{(2)}(\underset{\sim}{\underline{q}}_\alpha;\underset{\sim}{\underline{q}}_\beta;\underline{t})\right] + \underset{\sim}{\underline{F}}_{\alpha\beta,2}^{(2)}(\underset{\sim}{\underline{q}}_\alpha;\underset{\sim}{\underline{q}}_\beta;\underline{t})$$

$$\frac{d}{d\,t}\left[\underset{\sim}{\Lambda}_{\alpha\beta,1}^{(2)}(\underset{\sim}{\underline{q}}_\alpha;\underset{\sim}{\underline{q}}_\beta;\underline{t})\right] = -\nabla_{\underset{\sim}{\underline{R}}_\alpha}\left[\underline{kT}\ \ln\ \underline{C}_{\alpha\beta}^{(2)}(\underset{\sim}{\underline{q}}_\alpha;\underset{\sim}{\underline{q}}_\beta;\underline{t})\right] + \underset{\sim}{\underline{T}}_{\alpha\beta,1}^{(2)}(\underset{\sim}{\underline{q}}_\alpha;\underset{\sim}{\underline{q}}_\beta;\underline{t})$$

$$\frac{d}{d\,t}\left[\underset{\sim}{\Lambda}_{\alpha\beta,2}^{(2)}(\underset{\sim}{\underline{q}}_\alpha;\underset{\sim}{\underline{q}}_\beta;\underline{t})\right] = -\nabla_{\underset{\sim}{\underline{R}}_\beta}\left[\underline{kT}\ \ln\ \underline{C}_{\alpha\beta}^{(2)}(\underset{\sim}{\underline{q}}_\alpha;\underset{\sim}{\underline{q}}_\beta;\underline{t})\right] + \underset{\sim}{\underline{T}}_{\alpha\beta,2}^{(2)}(\underset{\sim}{\underline{q}}_\alpha;\underset{\sim}{\underline{q}}_\beta;\underline{t}).$$

Finally, from the equation of motion in $\alpha\beta$-pair position space we obtain the equations

$$\frac{d}{dt}\left[m_\alpha\, u_{\alpha\beta,1}^{(2)}(r_\alpha;r_\beta;t)\right] = -\nabla_{r_\alpha}\left[kT\ln C_{\alpha\beta}^{(2)}(r_\alpha;r_\beta;t)\right] + F_{\alpha\beta,1}^{(2)}(r_\alpha;r_\beta;t) \tag{68}$$

$$\frac{d}{dt}\left[m_\beta\, u_{\alpha\beta,2}^{(2)}(r_\alpha;r_\beta;t)\right] = -\nabla_{r_\beta}\left[kT\ln C_{\alpha\beta}^{(2)}(r_\alpha;r_\beta;t)\right] + F_{\alpha\beta,2}^{(2)}(r_\alpha;r_\beta;t).$$

At equilibrium, equations (65) and (67) become equations (35).

III. Continuum Model for a Fluid

We shall now consider our classical fluid system to be a dilute solution containing N solute molecules (or ions) dissolved in a solvent (molecules of species τ) so that the total volume of the system is V. There are N_α solute molecules of species α where α runs from 1 to σ. The solvent (τ) is assumed to be a viscous, incompressible, dielectric continuum; and the solute molecules are assumed to be rigid spheres, large in comparison to the solvent molecules, having rotational as well as translational degrees of freedom. We desire to obtain the steady state equations of motion for the solute molecules as they move in a solution which is placed in a constant external field and maintained at the constant uniform temperature T.

Steady State Equations of Motion

The forces which cause the solute molecules to move are: (1) forces due to external fields, (2) intermolecular forces due to the presence of neighboring solute molecules, i.e., the relaxation forces due to dissymmetries in the "atmospheres" of solute molecules about the reference solute molecules, (3) virtual forces arising from concentration gradients which together with thermal (random) agitations tend to produce uniform distributions, and (4) forces resulting from collisions with other solute molecules. Thus the average impressed force on a solute molecule of species α in the configuration $\mathbf{q}_\alpha$ is

$$\mathbf{F}_\alpha^{(1)}(\mathbf{q}_\alpha;t) - kT \nabla_{\mathbf{r}_\alpha} \ln C_\alpha^{(1)}(\mathbf{q}_\alpha;t) + \mathbf{K}_\alpha^{(1)}(\mathbf{q}_\alpha;t) . \tag{69}$$

Here, $\mathbf{F}_\alpha^{(1)}(\mathbf{q}_\alpha;t)$ is the average force on a solute molecule of species α, in configuration $\mathbf{q}_\alpha$, due to the external field and to molecular interactions with other solute molecules (see equation 27) ; the second term $-\underline{kT}\, \nabla_{\underset{\sim}{\underline{r}}_\alpha} \ln \underline{C}_\alpha^{(1)}(\underset{\sim}{\underline{q}}_\alpha;\underline{t})$ is the virtual force due to thermal motion, the temperature $\underline{T}$ being constant (see equation 65) ; and the last term $\underset{\sim}{\underline{K}}_\alpha^{(1)}(\underset{\sim}{\underline{q}}_\alpha;\underline{t})$ is the force due to collisions with other solute molecules. This kinetic effect[26] is equivalent to an asymmetry in the osmotic pressure on the reference solute molecule. Thus the osmotic pressure is treated as a virtual force acting on the individual molecules.[27]

As the solute molecules move through the viscous solvent they are subject to frictional resistance resulting, ultimately, from molecular impacts. The frictional force on a molecule will be taken as proportional to the linear velocity of that molecule. Then for the average frictional force on the molecule of species α in configuration $\underset{\sim}{\underline{q}}_\alpha$, we write

$$-\,\zeta_\alpha \left[\underset{\sim}{\underline{u}}_\alpha(\underset{\sim}{\underline{q}}_\alpha;\underline{t}) \;-\; \underset{\sim}{\underline{u}}\,{}_{|\alpha}(\underset{\sim}{\underline{r}}_\alpha\,|\,\underset{\sim}{\underline{q}}_\alpha;\underline{t})\right] \tag{70}$$

where $\underset{\sim}{\underline{u}}\,{}_{|\alpha}(\underset{\sim}{\underline{r}}_\alpha\,|\,\underset{\sim}{\underline{q}}_\alpha;\underline{t})$ is the mean local velocity of the fluid in the element of volume $\underline{d}^3\underset{\sim}{\underline{r}}_\alpha$ about $\underset{\sim}{\underline{r}}_\alpha$, there being a solute molecule of species α at the position $\underset{\sim}{\underline{r}}_\alpha$ in the orientation $\underset{\sim}{\underline{R}}_\alpha$; and where ζ_α is the coefficient of friction of a solute molecule of species α. This friction constant is a function of the dimensions of the molecules of species α and of the viscosity of the solvent. For our model consisting of rigid spheres moving in a viscous continuum, ξ_α is given by the formula due to Stokes

$$\xi_\alpha = 6\pi\eta\mathcal{R}_\alpha \tag{71}$$

where η is the viscosity of the solvent and where $\mathcal{R}_\alpha$ is the hydrodynamic radius of the sphere. As the impressed force on a solute molecule causes its velocity to increase the frictional force opposes this increase in direct proportion. Eventually a constant, steady state, velocity is reached such that the average total force (impressed plus frictional) on the solute molecule vanishes. Upon setting the average total force on the solute molecule of species α in configuration $\underset{\sim}{q}_\alpha$ to zero, and solving for $\underset{\sim}{u}_\alpha(\underset{\sim}{q}_\alpha)$, we obtain the <u>steady state equation of translational motion in α-configuration space</u>

$$\underset{\sim}{u}_\alpha(\underset{\sim}{q}_\alpha) = \underset{\sim}{u}_{|\alpha}(\underset{\sim}{r}_\alpha | \underset{\sim}{q}_\alpha) - D_\alpha \nabla_{\underset{\sim}{r}_\alpha} \ln C_\alpha^{(1)}(\underset{\sim}{q}_\alpha) + \frac{D_\alpha}{kT}\left[\underset{\sim}{F}_\alpha^{(1)}(\underset{\sim}{q}_\alpha) + \underset{\sim}{K}_\alpha^{(1)}(\underset{\sim}{q}_\alpha)\right] \tag{72}$$

in which

$$\underline{D}_{\alpha} = \frac{kT}{\xi_{\alpha}} = \frac{kT}{6\pi\eta \mathcal{R}_{\alpha}} \tag{73}$$

is the translational diffusion constant of the solute molecule of species α. This equation of motion is of the usual form, the factors causing the solute molecules to move being these: (1) flow of the solution as a whole, (2) diffusion resulting from thermal agitation, and obeying a diffusion law with the current density proportional to the gradient of the generic distribution function, and (3) forces on the solute molecules due to external fields and to other solute molecules. For a fluid at equilibrium in the absence of external forces the mean local velocities must vanish. Hence equation 72 reduces to

$$\underline{F}_{\alpha}^{(1,0)}(\underline{q}_{\alpha}) + \underline{K}_{\alpha}^{(1,0)}(\underline{q}_{\alpha}) - kT \nabla_{\underline{r}_{\alpha}} \ln \underline{C}_{\alpha}^{(1,0)}(\underline{q}_{\alpha}) = 0. \tag{74}$$

See equation 35.

In α-position space the steady state equation of translational motion is

$$\underline{u}_{\alpha}(\underline{r}_{\alpha}) = \underline{u}\,|_{\alpha}(\underline{r}_{\alpha} \,|\, \underline{r}_{\alpha}) - \underline{D}_{\alpha} \nabla_{\underline{r}_{\alpha}} \ln \underline{C}_{\alpha}^{(1)}(\underline{r}_{\alpha}) + \tag{75}$$

$$+ \frac{\underline{D}_\alpha}{\underline{kT}} \left[\underline{F}_\alpha^{(1)}(\underline{r}_\alpha) + \underline{K}_\alpha^{(1)}(\underline{r}_\alpha) \right]$$

a result easily obtained by multiplying equation 72 by $\underline{f}_\alpha^{(1)}(\underline{R}_\alpha | \underline{r}_\alpha)$ and averaging over orientations. For a fluid at equilibrium in the absence of external forces, this equation becomes

$$\underline{F}_\alpha^{(1,0)}(\underline{r}_\alpha) + \underline{K}_\alpha^{(1,0)}(\underline{r}_\alpha) = \underline{kT} \nabla_{\underline{r}_\alpha} \ln \underline{C}_\alpha^{(1,0)}(\underline{r}_\alpha) = 0. \tag{76}$$

To obtain the <u>steady state equation of rotational motion in α-configuration space</u> we require that the average total torque on a solute molecule of species α in configuration $\underline{q}_\alpha$ be zero. The average impressed torque on this molecule is

$$\underline{T}_\alpha^{(1)}(\underline{q}_\alpha;\underline{t}) - \underline{kT} \nabla_{\underline{R}_\alpha} \ln \underline{C}_\alpha^{(1)}(\underline{q}_\alpha;\underline{t}) \tag{77}$$

where $\underline{T}_\alpha^{(1)}(\underline{q}_\alpha;\underline{t})$ is the average torque on the molecule of species α due to the external field and to molecular interactions with other solute molecules (see equation 28); and where $-\underline{kT} \nabla_{\underline{R}_\alpha} \ln \underline{C}_\alpha^{(1)}(\underline{q}_\alpha;\underline{t})$ is the virtual torque due to thermal motion (see equation 65). The average frictional torque on the molecule of species α, assumed to be proportional to its angular velocity, is

$$- \zeta_{\alpha}^{(rot)} \left[\underset{\sim}{\Omega}_{\alpha}(\underset{\sim}{q}_{\alpha};t) - \underset{\sim}{\Omega}_{\alpha|\alpha}^{*}(\underset{\sim}{q}_{\alpha}|\underset{\sim}{q}_{\alpha};t) \right] \tag{78}$$

where $\underset{\sim}{\Omega}_{\alpha|\alpha}^{*}(\underset{\sim}{q}_{\alpha}|\underset{\sim}{q}_{\alpha};t)$ is the contribution to the average angular velocity of the molecule of species α at $\underset{\sim}{q}_{\alpha}$ due to the motion of the fluid in the element of volume $d^3\underset{\sim}{r}_{\alpha}$ about $\underset{\sim}{r}_{\alpha}$; and where $\zeta_{\alpha}^{(rot)}$ is the "inner frictional constant" or the rotational coefficient of friction of a solute molecule of species α. Clearly, this constant is a function of the size and shape of the molecule, and of the viscosity of the medium in which it is immersed. For a rigid sphere moving in a viscous continuum, $\zeta_{\alpha}^{(rot)}$ is given by the formula due to Stokes

$$\zeta_{\alpha}^{(rot)} = 8\pi\eta \, a_{\alpha}^{3} . \tag{79}$$

By equating the average total torque to zero, we obtain the desired equation

$$\underset{\sim}{\Omega}_{\alpha}(\underset{\sim}{q}_{\alpha}) = \underset{\sim}{\Omega}_{\alpha|\alpha}^{*}(\underset{\sim}{q}_{\alpha}|\underset{\sim}{q}_{\alpha}) - D_{\alpha}^{(rot)} \nabla_{\underset{\sim}{R}_{\alpha}} \ln c_{\alpha}^{(1)}(\underset{\sim}{q}_{\alpha}) + \tag{80}$$

$$+ \frac{D_\alpha^{(rot)}}{kT} \; \underset{\approx}{T}_\alpha^{(1)}(\underset{\approx}{q}_\alpha)$$

where

$$D_\alpha^{(rot)} = \frac{kT}{\zeta_\alpha^{(rot)}} = \frac{kT}{8\pi\eta \mathcal{R}_\alpha^3} \tag{81}$$

is the rotational diffusion constant of the solute molecule of species α. For a fluid at equilibrium in the absence of external fields, this equation of motion becomes

$$\underset{\approx}{T}_\alpha^{(1,0)}(\underset{\approx}{q}_\alpha) = kT \nabla_{\underset{\approx}{R}_\alpha} \ln C_\alpha^{(1,0)}(\underset{\approx}{q}_\alpha) = 0. \tag{82}$$

Analogously, the steady state equations of motion in $\alpha\beta$-configuration space are

$$\underset{\approx}{u}_{\alpha\beta,1}^{(2)}(\underset{\approx}{q}_\alpha;\underset{\approx}{q}_\beta) = \underset{\approx}{u}_{|\alpha\beta}(\underset{\approx}{r}_\alpha|\underset{\approx}{q}_\alpha;\underset{\approx}{q}_\beta) - D_\alpha \nabla_{\underset{\approx}{r}_\alpha} \ln C_{\alpha\beta}^{(2)}(\underset{\approx}{q}_\alpha;\underset{\approx}{q}_\beta) + \tag{83}$$

$$+ \frac{D_\alpha}{kT}\left[\underset{\approx}{F}_{\alpha\beta,1}^{(2)}(\underset{\approx}{q}_\alpha;\underset{\approx}{q}_\beta) + \underset{\approx}{K}_{\alpha\beta,1}^{(2)}(\underset{\approx}{q}_\alpha;\underset{\approx}{q}_\beta)\right]$$

$$\underline{u}_{\alpha\beta,2}^{(2)}(\underline{q}_\alpha;\underline{q}_\beta) = \underline{u}_{|\alpha\beta}(\underline{r}_\beta|\underline{q}_\alpha;\underline{q}_\beta) - D_\beta \nabla_{\underline{r}_\beta} \ln C_{\alpha\beta}^{(2)}(\underline{q}_\alpha;\underline{q}_\beta) +$$

$$+ \frac{D_\beta}{kT}\left[\underline{F}_{\alpha\beta,2}^{(2)}(\underline{q}_\alpha;\underline{q}_\beta) + \underline{K}_{\alpha\beta,2}^{(2)}(\underline{q}_\alpha;\underline{q}_\beta)\right]$$

$$\underline{\Omega}_{\alpha\beta,1}^{(2)}(\underline{q}_\alpha;\underline{q}_\beta) = \underline{\Omega}_{\alpha|\alpha\beta}^{*}(\underline{q}_\alpha|\underline{q}_\alpha;\underline{q}_\beta) +$$

$$- D_\alpha^{(\mathrm{rot})} \nabla_{\underline{R}_\alpha} \ln C_{\alpha\beta}^{(2)}(\underline{q}_\alpha;\underline{q}_\beta) + \frac{D_\alpha^{(\mathrm{rot})}}{kT} \underline{T}_{\alpha\beta,1}^{(2)}(\underline{q}_\alpha;\underline{q}_\beta)$$

$$\underline{\Omega}_{\alpha\beta,2}^{(2)}(\underline{q}_\alpha;\underline{q}_\beta) = \underline{\Omega}_{\beta|\alpha\beta}^{*}(\underline{q}_\beta|\underline{q}_\alpha;\underline{q}_\beta) +$$

$$- D_\beta^{(\mathrm{rot})} \nabla_{\underline{R}_\beta} \ln C_{\alpha\beta}^{(2)}(\underline{q}_\alpha;\underline{q}_\beta) + \frac{D_\beta^{(\mathrm{rot})}}{kT} \underline{T}_{\alpha\beta,2}^{(2)}(\underline{q}_\alpha;\underline{q}_\beta)$$

For a fluid at equilibrium in the absence of external fields, these equations reduce to

$$\underline{F}_{\alpha\beta,1}^{(2,0)}(\underline{q}_\alpha;\underline{q}_\beta) + \underline{K}_{\alpha\beta,1}^{(2,0)}(\underline{q}_\alpha;\underline{q}_\beta) = \tag{84}$$

$$= kT \nabla_{\underline{r}_\alpha} \ln C_{\alpha\beta}^{(2,0)}(\underline{q}_\alpha;\underline{q}_\beta)$$

$$\underline{\underline{F}}_{\alpha\beta,2}^{(2,0)}(\underline{q}_\alpha;\underline{q}_\beta) + \underline{\underline{K}}_{\alpha\beta,2}^{(2,0)}(\underline{q}_\alpha;\underline{q}_\beta) =$$

$$= \underline{kT} \nabla_{\underline{r}_\beta} \ln \underline{C}_{\alpha\beta}^{(2,0)}(\underline{q}_\alpha;\underline{q}_\beta)$$

$$\underline{\underline{T}}_{\alpha\beta,1}^{(2,0)}(\underline{q}_\alpha;\underline{q}_\beta) = \underline{kT} \nabla_{\underline{R}_\alpha} \ln \underline{C}_{\alpha\beta}^{(2,0)}(\underline{q}_\alpha;\underline{q}_\beta)$$

$$\underline{\underline{T}}_{\alpha\beta,2}^{(2,0)}(\underline{q}_\alpha;\underline{q}_\beta) = \underline{kT} \nabla_{\underline{R}_\beta} \ln \underline{C}_{\alpha\beta}^{(2,0)}(\underline{q}_\alpha;\underline{q}_\beta).$$

See equations 35.

Finally, the steady state equations of motion in $\alpha\beta$-position space are

(85)

$$\underline{u}_{\alpha\beta,1}^{(2)}(\underline{r}_\alpha;\underline{r}_\beta) = \underline{u}_{|\alpha\beta}(\underline{r}_\alpha|\underline{r}_\alpha;\underline{r}_\beta) - \underline{D}_\alpha \nabla_{\underline{r}_\alpha} \ln \underline{C}_{\alpha\beta}^{(2)}(\underline{r}_\alpha;\underline{r}_\beta) +$$

$$+ \frac{\underline{D}_\alpha}{\underline{kT}} \left[\underline{\underline{F}}_{\alpha\beta,1}^{(2)}(\underline{r}_\alpha;\underline{r}_\beta) + \underline{\underline{K}}_{\alpha\beta,1}^{(2)}(\underline{r}_\alpha;\underline{r}_\beta)\right]$$

$$\underline{u}_{\alpha\beta,2}^{(2)}(\underline{r}_\alpha;\underline{r}_\beta) = \underline{u}_{|\alpha\beta}(\underline{r}_\beta|\underline{r}_\alpha;\underline{r}_\beta) - \underline{D}_\beta \nabla_{\underline{r}_\beta} \ln \underline{C}_{\alpha\beta}^{(2)}(\underline{r}_\alpha;\underline{r}_\beta) +$$

$$+ \frac{\underline{D}_\beta}{\underline{kT}} \left[\underline{\underline{F}}_{\alpha\beta,2}^{(2)}(\underline{r}_\alpha;\underline{r}_\beta) + \underline{\underline{K}}_{\alpha\beta,2}^{(2)}(\underline{r}_\alpha;\underline{r}_\beta)\right] .$$

For a fluid at equilibrium in the absence of external fields, these equations reduce to

$$\underset{\sim}{F}_{\alpha\beta,1}^{(2,0)}(\underset{\sim}{r}_\alpha;\underset{\sim}{r}_\beta) + \underset{\sim}{K}_{\alpha\beta,1}^{(2,0)}(\underset{\sim}{r}_\alpha;\underset{\sim}{r}_\beta) = \tag{86}$$

$$= \underline{kT} \nabla_{\underset{\sim}{r}_\alpha} \ln \underline{C}_{\alpha\beta}^{(2,0)}(\underset{\sim}{r}_\alpha;\underset{\sim}{r}_\beta)$$

$$\underset{\sim}{F}_{\alpha\beta,2}^{(2,0)}(\underset{\sim}{r}_\alpha;\underset{\sim}{r}_\beta) + \underset{\sim}{K}_{\alpha\beta,2}^{(2,0)}(\underset{\sim}{r}_\alpha;\underset{\sim}{r}_\beta) =$$

$$= \underline{kT} \nabla_{\underset{\sim}{r}_\beta} \ln \underline{C}_{\alpha\beta}^{(2,0)}(\underset{\sim}{r}_\alpha;\underset{\sim}{r}_\beta) \quad .$$

We thus conclude our lengthy, but somewhat superficial, discussion of the equations of motion for rigid spheres in a viscous continuum. A more thorough approach, based possibly on the Liouville equation, yielding expressions for the kinetic effect and for the friction constants in terms of molecular forces would perhaps be preferable. However, equations of the form presented here, namely equations 64 and 85, were used by Onsager and Fuoss in their very successful theory of the electrical conductance of electrolytes.

Next, we shall combine these equations of motion with

the equations of continuity in order to obtain approximate equations of continuity which can be solved to yield expressions for the singlet and pair space generic distribution functions.

Approximate Steady State Equations of Continuity

For a steady state in a constant external field the secular change in $\underline{C}_\alpha^{(1)}(\underline{\underline{q}}_\alpha;\underline{t})$ vanishes, i.e., $\underline{C}_\alpha^{(1)}(\underline{\underline{q}}_\alpha;\underline{t})$ becomes independent of time, whence the equation of continuity in α-configuration space becomes

$$\nabla_{\underline{\underline{r}}_\alpha} \cdot \left[\underline{C}_\alpha^{(1)}(\underline{\underline{q}}_\alpha)\ \underline{\underline{u}}_\alpha(\underline{\underline{q}}_\alpha)\right] + \nabla_{\underline{\underline{R}}_\alpha} \cdot \left[\underline{C}_\alpha^{(1)}(\underline{\underline{q}}_\alpha)\ \Omega_\alpha(\underline{\underline{q}}_\alpha)\right] = 0. \tag{87}$$

We now substitute equations 72 and 80 in this equation and expand — employing the following sequence of operations:

(1) Write each of the following: $\underline{C}_\alpha^{(1)}(\underline{\underline{q}}_\alpha)$, $\underline{\underline{F}}_\alpha^{(1)}(\underline{\underline{q}}_\alpha)$, $\underline{\underline{K}}_\alpha^{(1)}(\underline{\underline{q}}_\alpha)$, and $\underline{\underline{T}}_\alpha^{(1)}(\underline{\underline{q}}_\alpha)$ as the sum of an equilibrium term of order zero in the external field plus a small perturbation term linear in the external field.

(2) Neglect streaming or flow of the solvent other than that caused by the external field. Both $\underline{\underline{u}}_{|\alpha}(\underline{\underline{r}}_\alpha|\underline{\underline{q}}_\alpha)$ and $\Omega_{\alpha|\alpha}^*(\underline{\underline{q}}_\alpha|\underline{\underline{q}}_\alpha)$, then, are linear in the external field.

(3) Neglect quadratic and higher terms in the external

field relative to terms linear in the external field.

(4) Utilize equations 74 and 82 whereby eight terms disappear by cancellation.

(5) Use the theorem for the divergence of the product of a scalar c and a vector $\mathbf{A}$, namely, $\nabla\cdot(c\mathbf{A}) = (\nabla c)\cdot\mathbf{A} + c\,\nabla\cdot\mathbf{A}$. Note that $c_\alpha^{(1,0)}(\mathbf{q}_\alpha)$, D_α, $D_\alpha^{(rot)}$, and T are constants, or rather, that we neglect their gradients; and since the solution is an incompressible fluid $\nabla_{\mathbf{r}_\alpha}\cdot\mathbf{u}_{|\alpha}(\mathbf{r}_\alpha|\mathbf{q}_\alpha) = 0$.

This yields, after rearrangement, the <u>approximate steady state equation of continuity in α-configuration space</u>

$$D_\alpha \nabla^2_{\mathbf{r}_\alpha} c_\alpha^{(1,1)}(\mathbf{q}_\alpha) + D_\alpha^{(rot)} \nabla^2_{\mathbf{R}_\alpha} c_\alpha^{(1,1)}(\mathbf{q}_\alpha) = \frac{D_\alpha}{kT} c_\alpha^{(1,0)}(\mathbf{q}_\alpha)\, \nabla_{\mathbf{r}_\alpha}\cdot\left[\mathbf{F}_\alpha^{(1,1)}(\mathbf{q}_\alpha) + \mathbf{K}_\alpha^{(1,1)}(\mathbf{q}_\alpha)\right] + \frac{D_\alpha^{(rot)}}{kT} c_\alpha^{(1,0)}(\mathbf{q}_\alpha)\, \nabla_{\mathbf{R}_\alpha}\cdot \mathbf{T}_\alpha^{(1,1)}(\mathbf{q}_\alpha) + c_\alpha^{(1,0)}(\mathbf{q}_\alpha)\, \nabla_{\mathbf{R}_\alpha}\cdot \boldsymbol{\Omega}_{\alpha|\alpha^*}(\mathbf{q}_\alpha|\mathbf{q}_\alpha) . \tag{88}$$

A similar procedure using equations 61, 75, and 76, gives

an <u>approximate steady state equation of continuity in</u> α-<u>position space</u>

$$\nabla^2_{\mathbf{r}_\alpha} \; c_\alpha^{(1,1)}(\mathbf{r}_\alpha) \; = \tag{89}$$

$$= \frac{1}{kT} \; c_\alpha^{(1,0)}(\mathbf{r}_\alpha) \; \nabla_{\mathbf{r}_\alpha} \cdot \left[\mathbf{F}_\alpha^{(1,1)}(\mathbf{r}_\alpha) + \mathbf{K}_\alpha^{(1,1)}(\mathbf{r}_\alpha) \right].$$

Incidently, for the special case of a homogeneous external field the perturbations become independent of absolute locations in the fluid. Then $c_\alpha^{(1,1)}(\mathbf{q}_\alpha)$ is a function of $\mathbf{R}_\alpha$ only; and $c_\alpha^{(1,1)}(\mathbf{r}_\alpha)$ is zero.

For a steady state in a constant external field $c_{\alpha\beta}^{(2)}(\mathbf{q}_\alpha;\mathbf{q}_\beta;t)$ is independent of time, and the equation of continuity in $\alpha\beta$-pair configuration space becomes

$$\nabla_{\mathbf{r}_\alpha} \cdot \left[c_{\alpha\beta}^{(2)}(\mathbf{q}_\alpha;\mathbf{q}_\beta) \; \mathbf{u}_{\alpha\beta,1}^{(2)}(\mathbf{q}_\alpha;\mathbf{q}_\beta) \right] + \tag{90}$$

$$+ \nabla_{\mathbf{r}_\beta} \cdot \left[c_{\alpha\beta}^{(2)}(\mathbf{q}_\alpha;\mathbf{q}_\beta) \; \mathbf{u}_{\alpha\beta,2}^{(2)}(\mathbf{q}_\alpha;\mathbf{q}_\beta) \right] +$$

$$+ \nabla_{\mathbf{R}_\alpha} \cdot \left[c_{\alpha\beta}^{(2)}(\mathbf{q}_\alpha;\mathbf{q}_\beta) \; \boldsymbol{\Omega}_{\alpha\beta,1}^{(2)}(\mathbf{q}_\alpha;\mathbf{q}_\beta) \right] +$$

$$+ \nabla_{\mathbf{R}_\beta} \cdot \left[c_{\alpha\beta}^{(2)}(\mathbf{q}_\alpha;\mathbf{q}_\beta) \; \boldsymbol{\Omega}_{\alpha\beta,2}^{(2)}(\mathbf{q}_\alpha;\mathbf{q}_\beta) \right] = 0 .$$

We now substitute equations 83 into this equation and expand. The following sequence of operations is employed:

(1) Write $\underline{C}_{\alpha\beta}^{(2)}(\underset{\sim}{q}_\alpha;\underset{\sim}{q}_\beta)$, $\underset{\sim}{F}_{\alpha\beta,1}^{(2)}(\underset{\sim}{q}_\alpha;\underset{\sim}{q}_\beta)$, etc., each as the sum of an equilibrium term of order zero in the external field plus a small perturbation term linear in the external field.

(2) Neglect streaming or flow of the solvent other than that caused by the external field. Then $\underset{\sim}{u}_{|\alpha\beta}(\underset{\sim}{r}_\alpha|\underset{\sim}{q}_\alpha;\underset{\sim}{q}_\beta)$, $\underset{\sim}{\Omega}_{\alpha|\alpha\beta}^{*}(\underset{\sim}{q}_\alpha|\underset{\sim}{q}_\alpha;\underset{\sim}{q}_\beta)$, etc., are terms linear in the external field.

(3) Neglect quadratic and higher terms in the external field relative to terms linear in the external field.

(4) Utilize equations 84 whereby ten terms disappear by cancellation.

(5) Use the theorem for the divergence of the product of a scalar and a vector. Note that $\underline{D}_\alpha$, $\underline{D}_\beta$, $\underline{D}_\alpha^{(rot)}$, $\underline{D}_\beta^{(rot)}$, and $\underline{T}$ are constants, or rather, that we neglect their gradients; and since the solution is an incompressible fluid, $\nabla_{\underset{\sim}{r}_\alpha} \cdot \underset{\sim}{u}_{|\alpha\beta}(\underset{\sim}{r}_\alpha|\underset{\sim}{q}_\alpha;\underset{\sim}{q}_\beta) = 0$ and $\nabla_{\underset{\sim}{r}_\beta} \cdot \underset{\sim}{u}_{|\alpha\beta}(\underset{\sim}{r}_\beta|\underset{\sim}{q}_\alpha;\underset{\sim}{q}_\beta) = 0$.

(6) Introduce relative coordinates $\underset{\sim}{r}_{\beta\alpha} = \underset{\sim}{r}_\beta - \underset{\sim}{r}_\alpha$ in position space. Since the equilibrium forces, torques, and distribution functions are independent of absolute locations they become functions only of $\underset{\sim}{R}_\alpha$, $\underset{\sim}{R}_\beta$, and $\underset{\sim}{r}_{\beta\alpha}$. Consequently, $\nabla_{\underset{\sim}{r}_{\beta\alpha}} = \nabla_{\underset{\sim}{r}_\beta} =$

$-\nabla_{\underline{r}_\alpha}$ when operating on these functions.

(7) Express $\underline{C}_{\alpha\beta}^{(2,0)}(\underline{q}_\alpha;\underline{q}_\beta)$ and $\underline{C}_{\alpha\beta}^{(2,1)}(\underline{q}_\alpha;\underline{q}_\beta)$ in terms of the pair correlation functions $\underline{g}_{\alpha\beta}^{(2,0)}(\underline{q}_\alpha;\underline{q}_\beta)$ and $\underline{g}_{\alpha\beta}^{(2,1)}(\underline{q}_\alpha;\underline{q}_\beta)$ defined by equations 30 and 41.

This gives a rather lengthy differential equation which we shall not reproduce here. For the special case of a homogeneous external field, a slight additional simplification is possible since the non-equilibrium perturbations to the forces, torques, distribution functions, etc., become independent of absolute locations in the fluid. They become functions only of $\underline{R}_\alpha$, $\underline{R}_\beta$, and $\underline{r}_{\beta\alpha}$ so that $\nabla_{\underline{r}_{\beta\alpha}} = \nabla_{\underline{r}_\beta} = -\nabla_{\underline{r}_\alpha}$ when operating on them. Our result is an <u>approximate steady state equation of continuity in $\alpha\beta$-pair configuration space</u>

$$(\underline{D}_\alpha + \underline{D}_\beta)\nabla^2_{\underline{r}_{\beta\alpha}}\underline{g}_{\alpha\beta}^{(2,1)} + \underline{D}_\alpha^{(rot)}\nabla^2_{\underline{R}_\alpha}\underline{g}_{\alpha\beta}^{(2,1)} + \tag{91}$$

$$\underline{D}_\beta^{(rot)}\nabla^2_{\underline{R}_\beta}\underline{g}_{\alpha\beta}^{(2,1)} + \left[\frac{\underline{D}_\alpha}{\underline{kT}}\left(\underline{F}_{\alpha\beta,1}^{(2,0)} + \underline{K}_{\alpha\beta,1}^{(2,0)}\right) + \right.$$

$$\left. - \frac{\underline{D}_\beta}{\underline{kT}}\left(\underline{F}_{\alpha\beta,2}^{(2,0)} + \underline{K}_{\alpha\beta,2}^{(2,0)}\right)\right] \cdot \nabla_{\underline{r}_{\beta\alpha}}\underline{g}_{\alpha\beta}^{(2,1)} +$$

$$- \frac{\underline{D}_\alpha^{(rot)}}{\underline{kT}}\underline{T}_{\alpha\beta,1}^{(2,0)} \cdot \nabla_{\underline{R}_\alpha}\underline{g}_{\alpha\beta}^{(2,1)} +$$

$$- \frac{\underline{D}_{\beta}^{(rot)}}{\underline{kT}} \underset{\approx}{T}_{\alpha\beta,2}^{(2,0)} \cdot \nabla_{\underset{\approx}{R}_{\beta}} \underline{g}_{\alpha\beta}^{(2,1)} +$$

$$\left[\frac{\underline{D}_{\alpha}}{\underline{kT}} \nabla_{\underset{\approx}{r}_{\beta\alpha}} \cdot \left(\underset{\approx}{F}_{\alpha\beta,1}^{(2,0)} + \underset{\approx}{K}_{\alpha\beta,1}^{(2,0)}\right) +\right.$$

$$- \frac{\underline{D}_{\beta}}{\underline{kT}} \nabla_{\underset{\approx}{r}_{\beta\alpha}} \cdot \left(\underset{\approx}{F}_{\alpha\beta,2}^{(2,0)} + \underset{\approx}{K}_{\alpha\beta,2}^{(2,0)}\right) +$$

$$- \frac{\underline{D}_{\alpha}^{(rot)}}{\underline{kT}} \nabla_{\underset{\approx}{R}_{\alpha}} \cdot \underset{\approx}{T}_{\alpha\beta,1}^{(2,0)} +$$

$$\left.- \frac{\underline{D}_{\beta}^{(rot)}}{\underline{kT}} \nabla_{\underset{\approx}{R}_{\beta}} \cdot \underset{\approx}{T}_{\alpha\beta,2}^{(2,0)}\right] \underline{g}_{\alpha\beta}^{(2,1)} +$$

$$\left[\frac{\underline{D}_{\alpha}}{\underline{kT}} \left(\underset{\approx}{F}_{\alpha\beta,1}^{(2,1)} + \underset{\approx}{K}_{\alpha\beta,1}^{(2,1)}\right) +\right.$$

$$\left.- \frac{\underline{D}_{\beta}}{\underline{kT}} \left(\underset{\approx}{F}_{\alpha\beta,2}^{(2,1)} + \underset{\approx}{K}_{\alpha\beta,2}^{(2,1)}\right)\right] \cdot \nabla_{\underset{\approx}{r}_{\beta\alpha}} \underline{g}_{\alpha\beta}^{(2,0)} +$$

$$- \frac{\underline{D}_{\alpha}^{(rot)}}{\underline{kT}} \underset{\approx}{T}_{\alpha\beta,1}^{(2,1)} \cdot \nabla_{\underset{\approx}{R}_{\alpha}} \underline{g}_{\alpha\beta}^{(2,0)} +$$

$$- \frac{\underline{D}_{\beta}^{(rot)}}{\underline{kT}} \underset{\approx}{T}_{\alpha\beta,2}^{(2,1)} \cdot \nabla_{\underset{\approx}{R}_{\beta}} \underline{g}_{\alpha\beta}^{(2,0)} +$$

$$\left[\frac{D_\alpha}{kT} \nabla_{\underset{\sim}{r}_{\beta\alpha}} \cdot \left(\underset{\sim}{F}_{\alpha\beta,1}{}^{(2,1)} + \underset{\sim}{K}_{\alpha\beta,1}{}^{(2,1)}\right) + \right.$$

$$- \frac{D_\beta}{kT} \nabla_{\underset{\sim}{r}_{\beta\alpha}} \cdot \left(\underset{\sim}{F}_{\alpha\beta,2}{}^{(2,1)} + \underset{\sim}{K}_{\alpha\beta,2}{}^{(2,1)}\right) +$$

$$- \frac{D_\alpha{}^{(rot)}}{kT} \nabla_{\underset{\sim}{R}_\alpha} \cdot \underset{\sim}{T}_{\alpha\beta,1}{}^{(2,1)} +$$

$$\left. - \frac{D_\beta{}^{(rot)}}{kT} \nabla_{\underset{\sim}{R}_\beta} \cdot \underset{\sim}{T}_{\alpha\beta,2}{}^{(2,1)}\right] g_{\alpha\beta}{}^{(2,0)} =$$

$$= \left[\underset{\sim}{u}\,|\alpha\beta(\underset{\sim}{r}_\beta) - \underset{\sim}{u}\,|\alpha\beta(\underset{\sim}{r}_\alpha)\right] \cdot \nabla_{\underset{\sim}{r}_{\beta\alpha}} g_{\alpha\beta}{}^{(2,0)} +$$

$$+ \underset{\sim}{\Omega}_{\alpha|\alpha\beta}{}^{*} \cdot \nabla_{\underset{\sim}{R}_\alpha} g_{\alpha\beta}{}^{(2,0)} +$$

$$+ \underset{\sim}{\Omega}_{\beta|\alpha\beta}{}^{*} \cdot \nabla_{\underset{\sim}{R}_\beta} g_{\alpha\beta}{}^{(2,0)} +$$

$$+ g_{\alpha\beta}{}^{(2,0)} \nabla_{\underset{\sim}{R}_\alpha} \cdot \underset{\sim}{\Omega}_{\alpha|\alpha\beta}{}^{*} +$$

$$+ g_{\alpha\beta}{}^{(2,0)} \nabla_{\underset{\sim}{R}_\beta} \cdot \underset{\sim}{\Omega}_{\beta|\alpha\beta}{}^{*} \,.$$

In order to obtain an approximate expression for $\underline{g}_{\alpha\beta}^{(2,1)}(\underset{\sim}{q}_\alpha;\underset{\sim}{q}_\beta)$, we must integrate this formidable equation. Fortunately, in the cases which we shall consider additional approximations are possible.

The boundary conditions on $\underline{g}_{\alpha\beta}^{(2,1)}(\underset{\sim}{q}_\alpha;\underset{\sim}{q}_\beta)$ are of course: (1) no correlation at infinite separation of molecules in the fluid — equation 46 — with expressions for $\underline{C}_\alpha^{(1,1)}(\underset{\sim}{q}_\alpha)$ and $\underline{C}_\beta^{(1,1)}(\underset{\sim}{q}_\beta)$, which are independent of $\underset{\sim}{r}_\alpha$ and $\underset{\sim}{r}_\beta$ for the special case of a homogeneous external field, obtained by solving the singlet space equation of continuity, equation 88; and (2) no interpenetration of rigid spheres — equation 26 — which when expanded using the same approximations used in obtaining equation 91 becomes:

$$\Bigg[(\underline{D}_\alpha + \underline{D}_\beta)\nabla_{\underset{\sim}{r}_{\beta\alpha}}\underline{g}_{\alpha\beta}^{(2,1)} + \Bigg[\frac{\underline{D}_\alpha}{\underline{kT}}\left(\underset{\sim}{F}_{\alpha\beta,1}^{(2,0)} + \underset{\sim}{K}_{\alpha\beta,1}^{(2,0)}\right) + {} - \frac{\underline{D}_\beta}{\underline{kT}}\left(\underset{\sim}{F}_{\alpha\beta,2}^{(2,0)} + \underset{\sim}{K}_{\alpha\beta,2}^{(2,0)}\right)\Bigg]\underline{g}_{\alpha\beta}^{(2,1)} + {} + \Bigg[\frac{\underline{D}_\alpha}{\underline{kT}}\left(\underset{\sim}{F}_{\alpha\beta,1}^{(2,1)} + \underset{\sim}{K}_{\alpha\beta,1}^{(2,1)}\right) - \frac{\underline{D}_\beta}{\underline{kT}}\left(\underset{\sim}{F}_{\alpha\beta,2}^{(2,1)} + {} + \underset{\sim}{K}_{\alpha\beta,2}^{(2,1)}\right)\Bigg]\underline{g}_{\alpha\beta}^{(2,0)} + {} + \left[\underset{\sim}{u}\,|_{\alpha\beta}(\underset{\sim}{r}_\alpha) - \underset{\sim}{u}\,|_{\alpha\beta}(\underset{\sim}{r}_\beta)\right]\underline{g}_{\alpha\beta}^{(2,0)}\Bigg]_{\underline{r}_{\alpha\beta} = \underline{a}_{\alpha\beta}} \cdot \underset{\sim}{1}_{\underline{r}_{\beta\alpha}} = 0. \tag{92}$$

Using equations 64, 85, and 86, an analogous procedure gives an <u>approximate steady state equation of continuity in</u> <u>$\alpha\beta$-pair position space</u>

(93)

$$
\begin{aligned}
&(D_\alpha + D_\beta)\, \nabla^2_{\underset{\approx}{r}_{\beta\alpha}}\, g_{\alpha\beta}^{(2,1)} + \left[\frac{D_\alpha}{kT}\left(\underset{\approx}{F}_{\alpha\beta,1}^{(2,0)} + \underset{\approx}{K}_{\alpha\beta,1}^{(2,0)}\right) + \right.\\
&\left. - \frac{D_\beta}{kT}\left(\underset{\approx}{F}_{\alpha\beta,2}^{(2,0)} + \underset{\approx}{K}_{\alpha\beta,2}^{(2,0)}\right)\right] \cdot \nabla_{\underset{\approx}{r}_{\beta\alpha}}\, g_{\alpha\beta}^{(2,1)} + \\
&+ \left[\frac{D_\alpha}{kT}\, \nabla_{\underset{\approx}{r}_{\beta\alpha}} \cdot \left(\underset{\approx}{F}_{\alpha\beta,1}^{(2,0)} + \underset{\approx}{K}_{\alpha\beta,1}^{(2,0)}\right) + \right.\\
&\left. - \frac{D_\beta}{kT}\, \nabla_{\underset{\approx}{r}_{\beta\alpha}} \cdot \left(\underset{\approx}{F}_{\alpha\beta,2}^{(2,0)} + \underset{\approx}{K}_{\alpha\beta,2}^{(2,0)}\right)\right] g_{\alpha\beta}^{(2,1)} + \\
&+ \left[\frac{D_\alpha}{kT}\left(\underset{\approx}{F}_{\alpha\beta,1}^{(2,1)} + \underset{\approx}{K}_{\alpha\beta,1}^{(2,1)}\right) + \right.\\
&\left. - \frac{D_\beta}{kT}\left(\underset{\approx}{F}_{\alpha\beta,2}^{(2,1)} + \underset{\approx}{K}_{\alpha\beta,2}^{(2,1)}\right)\right] \cdot \nabla_{\underset{\approx}{r}_{\beta\alpha}}\, g_{\alpha\beta}^{(2,0)} + \\
&+ \left[\frac{D_\alpha}{kT}\, \nabla_{\underset{\approx}{r}_{\beta\alpha}} \cdot \left(\underset{\approx}{F}_{\alpha\beta,1}^{(2,1)} + \underset{\approx}{K}_{\alpha\beta,1}^{(2,1)}\right) + \right.\\
&\left. - \frac{D_\beta}{kT}\, \nabla_{\underset{\approx}{r}_{\beta\alpha}} \cdot \left(\underset{\approx}{F}_{\alpha\beta,2}^{(2,1)} + \underset{\approx}{K}_{\alpha\beta,2}^{(2,1)}\right)\right] g_{\alpha\beta}^{(2,0)} = \\
&= \left[\underset{\approx}{u}\,|_{\alpha\beta}(\underset{\approx}{r}_\beta) - \underset{\approx}{u}\,|_{\alpha\beta}(\underset{\approx}{r}_\alpha)\right] \cdot \nabla_{\underset{\approx}{r}_{\beta\alpha}}\, g_{\alpha\beta}^{(2,0)} .
\end{aligned}
$$

In order to obtain an approximate expression for $\underline{g}_{\alpha\beta}^{(2,1)}(\underline{r}_\alpha;\underline{r}_\beta)$, we must solve this equation.

The boundary conditions on $\underline{g}_{\alpha\beta}^{(2,1)}(\underline{r}_\alpha;\underline{r}_\beta)$ are again: (1) no correlation at infinite separation of molecules in the fluid, which becomes

$$\lim_{\underline{r}_{\alpha\beta}\to+\infty} \underline{g}_{\alpha\beta}^{(2,1)}(\underline{r}_\alpha;\underline{r}_\beta) = 0 \tag{94}$$

for the special case of a homogeneous external field; and (2) no interpenetration of rigid spheres, as given by an equation similar to equation 26

$$\left(\underline{c}_{\alpha\beta}^{(2)}(\underline{r}_\alpha;\underline{r}_\beta)\left[\underline{u}_{\alpha\beta,1}^{(2)}(\underline{r}_\alpha;\underline{r}_\beta) - \underline{u}_{\alpha\beta,2}^{(2)}(\underline{r}_\alpha;\underline{r}_\beta)\right]\right)_{\underline{r}_{\alpha\beta}=\underline{a}_{\alpha\beta}} \cdot \underline{1}_{\underline{r}_{\beta\alpha}} = 0 \tag{95}$$

which appears the same as equation 92 when expanded using the same approximations used in obtaining equation 93.

We have seen that in order to obtain approximate expressions for the non-equilibrium perturbations $\underline{c}_\alpha^{(1,1)}(\underline{q}_\alpha)$, $\underline{c}_\alpha^{(1,1)}(\underline{r}_\alpha)$, $\underline{g}_{\alpha\beta}^{(2,1)}(\underline{q}_\alpha;\underline{q}_\beta)$, and $\underline{g}_{\alpha\beta}^{(2,1)}(\underline{r}_\alpha;\underline{r}_\beta)$ we must integrate equations 88, 89, 91, and 93 respectively, subject to the appropriate boundary conditions. Before this can be accomplished, however, we must have usable expressions for the forces, torques, and velocities appearing in these equations. Thus in the remaining paragraphs of this section the movements of solute molecules resulting from assymmetries in the osmotic pressure and from flow of the fluid as a whole will be considered.

The Kinetic Effect

Our discussion will be based on the Navier-Stokes equation of motion

$$\rho \frac{d\mathbf{u}}{dt} = -\nabla_{\mathbf{r}} P + (\eta/3 + \varphi)\nabla_{\mathbf{r}}(\nabla_{\mathbf{r}} \cdot \mathbf{u}) + \eta \nabla_{\mathbf{r}}^2 \mathbf{u} + \mathbf{X} \tag{96}$$

which can be obtained from the general equation of change.[28] Here η is the shear viscosity, φ is the bulk viscosity, P is the presure, and $\mathbf{X}$ is the volume force. The pressure P includes the osmotic pressure of the solution as well as the hydrostatic pressure of the pure solvent. For steady-state motion of a viscous non-compressible fluid this equation becomes

$$\eta \nabla_{\mathbf{r}}^2 \mathbf{u}(\mathbf{r}) = \nabla_{\mathbf{r}} P(\mathbf{r}) - \mathbf{X}(\mathbf{r}) \tag{97}$$

with the vector $\underset{\sim}{r}$ introduced so as to note explicitly the dependence of $\underset{\sim}{u}$, $\underline{P}$, and $\underset{\sim}{X}$ on position. The volume force on the fluid at position $\underset{\sim}{r}$ is given by

$$\underset{\sim}{X}(\underset{\sim}{r}) = \sum_{\alpha=1}^{\sigma} \underset{\sim}{F}_{\alpha}^{(1)}(\underset{\sim}{r})\ \underline{C}_{\alpha}^{(1)}(\underset{\sim}{r}).$$

We now consider the motion of the fluid in the vicinity of a reference molecule of species α in orientation $\underset{\sim}{R}_{\alpha}$ at position $\underset{\sim}{r}_{\alpha}$ in a constant homogeneous external field. The Navier-Stokes equation of motion becomes

$$\eta \nabla_{\underset{\sim}{r}}^{2}\ \underset{\sim}{u}_{|\alpha}(\underset{\sim}{r}|\underset{\sim}{q}_{\alpha}) = \nabla_{\underset{\sim}{r}}\ \underline{P}_{|\alpha}(\underset{\sim}{r}|\underset{\sim}{q}_{\alpha}) - \underset{\sim}{X}_{|\alpha}(\underset{\sim}{r}|\underset{\sim}{q}_{\alpha}) \tag{98}$$

where $\underset{\sim}{u}_{|\alpha}(\underset{\sim}{r}|\underset{\sim}{q}_{\alpha})$ is the mean local velocity of the fluid at $\underset{\sim}{r}$, there being a molecule of species α at $\underset{\sim}{q}_{\alpha}$; where $\underline{P}_{|\alpha}(\underset{\sim}{r}|\underset{\sim}{q}_{\alpha})$ is the pressure of the fluid at $\underset{\sim}{r}$, there being a molecule of species α at $\underset{\sim}{q}_{\alpha}$; and where

$$\underset{\sim}{X}_{|\alpha}(\underset{\sim}{r}|\underset{\sim}{q}_{\alpha}) = \sum_{\beta=1}^{\sigma} \underset{\sim}{F}_{\alpha\beta,2}^{(2)}(\underset{\sim}{q}_{\alpha};\underset{\sim}{r})\ \underline{C}_{\beta|\alpha}^{(2)}(\underset{\sim}{r}|\underset{\sim}{q}_{\alpha})$$

is the volume force on the fluid at $\underset{\sim}{r}$, there being a molecule of species α at $\underset{\sim}{q}_\alpha$. Effects due to the reference molecule must vanish at large distances from it, whence

$$\lim_{\underline{r}_{\alpha\beta} \to +\infty} \underset{\sim}{u}_{|\alpha}(\underset{\sim}{r}_\beta | \underset{\sim}{q}_\alpha) = \underset{\sim}{u}(\underset{\sim}{r}_\beta)$$

$$\lim_{\underline{r}_{\alpha\beta} \to +\infty} \underline{P}_{|\alpha}(\underset{\sim}{r}_\beta | \underset{\sim}{q}_\alpha) = \underline{P}(\underset{\sim}{r}_\beta)$$

$$\lim_{\underline{r}_{\alpha\beta} \to +\infty} \underset{\sim}{X}_{|\alpha}(\underset{\sim}{r}_\beta | \underset{\sim}{q}_\alpha) = \underset{\sim}{X}(\underset{\sim}{r}_\beta)$$

so that equation 98 reduces to equation 97. Subtracting equation 97 from equation 98, we obtain the result

$$\eta \nabla^2_{\underset{\sim}{r}} \underset{\sim}{v}_{|\alpha}(\underset{\sim}{r} | \underset{\sim}{q}_\alpha) = \nabla_{\underset{\sim}{r}} \pi_{|\alpha}(\underset{\sim}{r} | \underset{\sim}{q}_\alpha) - \underset{\sim}{Y}_{|\alpha}(\underset{\sim}{r} | \underset{\sim}{q}_\alpha) \tag{99}$$

where, by definition,

$$\underset{\sim}{v}_{|\alpha}(\underset{\sim}{r} | \underset{\sim}{q}_\alpha) = \underset{\sim}{u}_{|\alpha}(\underset{\sim}{r} | \underset{\sim}{q}_\alpha) - \underset{\sim}{u}(\underset{\sim}{r}) \tag{100}$$

$$\pi_{|\alpha}(\underset{\sim}{r} | \underset{\sim}{q}_\alpha) = \underline{P}_{|\alpha}(\underset{\sim}{r} | \underset{\sim}{q}_\alpha) - \underline{P}(\underset{\sim}{r})$$

$$\underset{\approx}{Y}_{|\alpha}(\underset{\approx}{r}|\underset{\approx}{q}_\alpha) = \underset{\approx}{X}_{|\alpha}(\underset{\approx}{r}|\underset{\approx}{q}_\alpha) - \underset{\approx}{X}(\underset{\approx}{r}).$$

Given a suitable expression for $\underset{\approx}{Y}_{|\alpha}(\underset{\approx}{r}|\underset{\approx}{q}_\alpha)$, we can integrate equation 99 and obtain values for $\underset{\approx}{v}_{|\alpha}(\underset{\approx}{r}|\underset{\approx}{q}_\alpha)$ and for $\pi_{|\alpha}(\underset{\approx}{r}|\underset{\approx}{q}_\alpha)$. The kinetic force $\underset{\approx}{K}_\alpha^{(1)}(\underset{\approx}{q}_\alpha)$ arising from an asymmetry in the pressure on the spherically symmetrical reference molecule of species α is obtained by integrating the inwardly directed normal pressure $-\pi_{|\alpha}(\underset{\approx}{r}_\alpha + \mathcal{R}_\alpha \underset{\approx}{1}_{\underline{r}_{\beta\alpha}}|\underset{\approx}{q}_\alpha)\, \underset{\approx}{1}_{\underline{r}_{\beta\alpha}}$ over the hydrodynamical surface $\underline{r}_{\beta\alpha} = \mathcal{R}_\alpha$ of the reference molecule, i.e.,

$$\underset{\approx}{K}_\alpha^{(1)}(\underset{\approx}{q}_\alpha) = -\iint \pi_{|\alpha}(\underset{\approx}{r}_\alpha + \mathcal{R}_\alpha \underset{\approx}{1}_{\underline{r}_{\beta\alpha}}|\underset{\approx}{q}_\alpha)\, \underset{\approx}{1}_{\underline{r}_{\beta\alpha}}\, \mathcal{R}_\alpha^2 \sin \theta_{\beta\alpha}\, \underline{d}\theta_{\beta\alpha}\, \underline{d}\varphi_{\beta\alpha}. \tag{101}$$

Unfortunately, solving equation 99 is often a difficult task. On the brighter side, however, Onsager and Fuoss[29] have developed a useful alternative procedure to which we now turn.

Since $\nabla_{\underset{\approx}{r}} \cdot \underset{\approx}{v}_{|\alpha}(\underset{\approx}{r}|\underset{\approx}{q}_\alpha)$ vanishes, equation 99 becomes

$$\eta \nabla_{\underset{\approx}{r}} \times \nabla_{\underset{\approx}{r}} \times \underset{\approx}{v}_{|\alpha}(\underset{\approx}{r}|\underset{\approx}{q}_\alpha) = \underset{\approx}{Y}_{|\alpha}(\underset{\approx}{r}|\underset{\approx}{q}_\alpha) - \nabla_{\underset{\approx}{r}} \pi_{|\alpha}(\underset{\approx}{r}|\underset{\approx}{q}_\alpha).$$

Operating on this equation with $\nabla_{\underline{\underline{r}}}\cdot$, we obtain the equation

$$\nabla_{\underline{\underline{r}}}^{2}\ \pi_{|\alpha}(\underline{\underline{r}}|\underline{\underline{q}}_\alpha) = \nabla_{\underline{\underline{r}}}\cdot\ \underline{\underline{Y}}_{|\alpha}(\underline{\underline{r}}|\underline{\underline{q}}_\alpha).$$

With $\underline{\underline{F}}_{\alpha\beta,2}^{(2)}(\underline{\underline{q}}_\alpha;\underline{\underline{r}})$ given by $+\ \underline{kT}\nabla_{\underline{\underline{r}}}\ \ln\ \underline{C}_{\alpha\beta}^{(2)}(\underline{\underline{q}}_\alpha;\underline{\underline{r}})$ and with $\underline{\underline{F}}_{\beta}^{(1)}(\underline{\underline{r}})$ given by $+\ \underline{kT}\nabla_{\underline{\underline{r}}}\ \ln\ \underline{C}_{\beta}^{(1)}(\underline{\underline{r}})$, see equations 67 and 66, the volume force $\underline{\underline{Y}}_{|\alpha}(\underline{\underline{r}}|\underline{\underline{q}}_\alpha)$ is

$$\underline{\underline{Y}}_{|\alpha}(\underline{\underline{r}}|\underline{\underline{q}}_\alpha) = \underline{kT}\nabla_{\underline{\underline{r}}} \sum_{\beta=1}^{\sigma} \left[\underline{C}_{\beta|\alpha}^{(2)}(\underline{\underline{r}}|\underline{\underline{q}}_\alpha) - \underline{C}_{\beta}^{(1)}(\underline{\underline{r}})\right]$$

whence

$$\pi_{|\alpha}(\underline{\underline{r}}|\underline{\underline{q}}_\alpha) = \underline{kT}\sum_{\beta=1}^{\sigma} \underline{C}_{\beta}^{(1)}(\underline{\underline{r}}) \left[\underline{G}_{\alpha\beta}^{(2)}(\underline{\underline{q}}_\alpha;\underline{\underline{r}}) - 1\right].$$

For a homogeneous external field,

$$\pi_{|\alpha}(\underline{\underline{r}}|\underline{\underline{q}}_\alpha) = \underline{kT}\sum_{\beta=1}^{\sigma} \underline{C}_{\beta} \left[\underline{G}_{\alpha\beta}^{(2)}(\underline{\underline{q}}_\alpha;\underline{\underline{r}}) - 1\right],$$

and for the kinetic force $\underline{\underline{K}}_{\alpha}^{(1)}(\underline{\underline{q}}_\alpha)$ we have the expression

$$\underline{\underset{\sim}{K}}_\alpha^{(1)}(\underset{\sim}{\underline{q}}_\alpha) = \tag{102}$$

$$= -\underline{kT} \sum_{\beta=1}^{\sigma} \underline{c}_\beta \iint \underline{G}_{\alpha\beta}^{(2)}(\underset{\sim}{\underline{q}}_\alpha ; \underset{\sim}{\underline{r}}_\alpha + \mathcal{R}_\alpha \underset{\sim}{\underline{1}}_{\underline{r}_{\beta\alpha}}) \, \underset{\sim}{\underline{1}}_{\underline{r}_{\beta\alpha}} \mathcal{R}_\alpha^2 \sin \theta_{\beta\alpha} \, \underline{d}\theta_{\beta\alpha} \, \underline{d}\varphi_{\beta\alpha} .$$

This effect is linear in the concentration to a first approximation.

To obtain an expression for $\underline{\underset{\sim}{K}}_\alpha^{(1)}(\underset{\sim}{\underline{r}}_\alpha)$ we multiply $\underline{\underset{\sim}{K}}_\alpha^{(1)}(\underset{\sim}{\underline{q}}_\alpha)$ by $\underline{f}_\alpha^{(1)}(\underset{\sim}{\underline{R}}_\alpha | \underset{\sim}{\underline{r}}_\alpha)$ and average over orientations, i.e.,

$$\underline{\underset{\sim}{K}}_\alpha^{(1)}(\underset{\sim}{\underline{r}}_\alpha) = \int \underline{f}_\alpha^{(1)}(\underset{\sim}{\underline{R}}_\alpha | \underset{\sim}{\underline{r}}_\alpha) \, \underline{\underset{\sim}{K}}_\alpha^{(1)}(\underset{\sim}{\underline{q}}_\alpha) \, \underline{d}^3 \underset{\sim}{\underline{R}}_\alpha .$$

For a homogeneous external field, we have

$$\pi_{|\alpha}(\underset{\sim}{\underline{r}} | \underset{\sim}{\underline{r}}_\alpha) = \underline{kT} \sum_{\beta=1}^{\sigma} \underline{c}_\beta \left[\underline{g}_{\alpha\beta}^{(2)}(\underset{\sim}{\underline{r}}_\alpha ; \underset{\sim}{\underline{r}}) - 1 \right]$$

whence

$$\underline{\underset{\sim}{K}}_\alpha^{(1)}(\underset{\sim}{\underline{r}}_\alpha) = \tag{103}$$

$$= -\underline{kT} \sum_{\beta=1}^{\sigma} \underline{c}_\beta \iint \underline{g}_{\alpha\beta}^{(2)}(\underset{\sim}{\underline{r}}_\alpha ; \underset{\sim}{\underline{r}}_\alpha + \mathcal{R}_\alpha \underset{\sim}{\underline{1}}_{\underline{r}_{\beta\alpha}}) \, \underset{\sim}{\underline{1}}_{\underline{r}_{\beta\alpha}} \mathcal{R}_\alpha^2 \sin \theta_{\beta\alpha} \, \underline{d}\theta_{\beta\alpha} \, \underline{d}\varphi_{\beta\alpha} .$$

The Electrophoretic Effect

When a solute molecule or ion moves through a viscous medium it tends to drag along with it the solution in its vicinity. This effect extends for relatively great distances. At equilibrium, in the absence of external force fields, the solute particles move in a random manner. In the presence of an external field, however, they may be transported through the solution. As a result the fluid as a whole may be set into motion — a motion described by the macroscopic stream velocity $\underline{u}(\underline{r};\underline{t})$. For steady state motion in a constant homogeneous external field, $\underline{u}$ is a constant independent of both position and time. We are not interested in evaluating $\underline{u}$; rather, we wish to know whether or not the fluid motion in the neighborhood of a solute molecule, say of species α, imparts to that molecule a velocity different from $\underline{u}$, i.e., we seek to obtain an expression for $\underline{v}_{|\alpha}(\underline{r}_\alpha|\underline{q}_\alpha)$ as defined by equation 100.

The external force acting on an element of volume $\underline{d}^3\underline{r}_\beta$ near a molecule of species α at $\underline{q}_\alpha$, over and above the external force that would be acting on that volume element if the molecule of species α were not necessarily at $\underline{q}_\alpha$, is given by

$$\underline{d}\underline{F}_{\underline{v}}^{(e)} = \underline{Y}_{|\alpha}^{(e)}(\underline{r}_\beta|\underline{q}_\alpha) \quad \underline{d}^3\underline{r}_\beta \tag{104}$$

where

$$\underset{\approx}{Y}_{|\alpha}^{(e)}(\underset{\approx}{r}_\beta|\underset{\approx}{q}_\alpha) = \underset{\approx}{X}_{|\alpha}^{(e)}(\underset{\approx}{r}_\beta|\underset{\approx}{q}_\alpha) - \underset{\approx}{X}^{(e)}(\underset{\approx}{r}_\beta)$$

with

$$\underset{\approx}{X}^{(e)}(\underset{\approx}{r}_\beta) = \sum_{\gamma=1}^{\sigma} \underset{\approx}{F}_\gamma^{(e)}(\underset{\approx}{r}_\beta) \; \underline{C}_\gamma^{(1)}(\underset{\approx}{r}_\beta)$$

$$\underset{\approx}{X}_{|\alpha}^{(e)}(\underset{\approx}{r}_\beta|\underset{\approx}{q}_\alpha) = \sum_{\gamma=1}^{\sigma} \underset{\approx}{F}_{\gamma|\alpha}^{(e)}(\underset{\approx}{r}_\beta|\underset{\approx}{q}_\alpha) \; \underline{C}_{\gamma|\alpha}^{(2)}(\underset{\approx}{r}_\beta|\underset{\approx}{q}_\alpha) \; .$$

Upon introducing relative coordinates $\underset{\approx}{r}_{\beta\alpha} = \underset{\approx}{r}_\beta - \underset{\approx}{r}_\alpha$ so that $\underline{d}^3\underset{\approx}{r}_{\beta\alpha} = \underline{d}^3\underset{\approx}{r}_\beta$ and $\underset{\approx}{Y}_{|\alpha}^{(e)}(\underset{\approx}{r}_\beta|\underset{\approx}{q}_\alpha)$ becomes a function of $\underset{\approx}{r}_{\beta\alpha}$ and $\underset{\approx}{q}_\alpha$ which is indicated by writting $\underset{\approx}{Y}_{|\alpha}^{(e)}(\underset{\approx}{r}_{\beta\alpha}|\underset{\approx}{q}_\alpha)$, we can write

$$\underline{d}\,\underset{\approx}{F}_{\underline{v}}^{(e)} = \underset{\approx}{Y}_{|\alpha}^{(e)}(\underset{\approx}{r}_{\beta\alpha}|\underset{\approx}{q}_\alpha) \; \underline{d}^3\,\underset{\approx}{r}_{\beta\alpha} \; .$$

By integrating over angles, we obtain an expression for the external force

$$\underline{d}\,\underset{\approx}{F}_{\underline{s}}^{(e)} = \int_{\varphi_{\beta\alpha}=0}^{\varphi_{\beta\alpha}=2\pi} \int_{\theta_{\beta\alpha}=0}^{\theta_{\beta\alpha}=\pi} \underset{\approx}{Y}_{|\alpha}^{(e)}(\underset{\approx}{r}_{\beta\alpha}|\underset{\approx}{q}_\alpha) \; \underline{d}^3\underset{\approx}{r}_{\beta\alpha} \qquad (105)$$

acting on a spherical shell of thickness $d\,r_{\beta\alpha}$ around the reference molecule of species α at $\underset{\sim}{q}_\alpha$. Now if this force is distributed uniformly over the shell, and if it is in the direction of the applied external force; Stokes' law is applicable. Accordingly, the contents of the shell move with velocity $d\,\underset{\sim}{F}_s^{(e)}/\,6\pi\eta\,r_{\beta\alpha}$. The electrophoretic velocity $\underset{\sim}{v}_{|\alpha}(\underset{\sim}{r}_\alpha|\underset{\sim}{q}_\alpha)$ at the site $\underset{\sim}{r}_\alpha$ of the reference molecule is obtained by summing the contributions due to all shells, i.e., by integrating over all values of $r_{\beta\alpha}$ from $\mathcal{R}_\alpha$ to $+\infty$, with the result

$$\underset{\sim}{v}_{|\alpha}(\underset{\sim}{r}_\alpha|\underset{\sim}{q}_\alpha) = \frac{1}{6\pi\eta}\int_{r_{\beta\alpha}=\mathcal{R}_\alpha}^{r_{\beta\alpha}=+\infty} \frac{1}{r_{\beta\alpha}}\,\underset{\sim}{Y}_{|\alpha}^{(e)}(\underset{\sim}{r}_{\beta\alpha}|\underset{\sim}{q}_\alpha)\,d^3\underset{\sim}{r}_{\beta\alpha}. \tag{106}$$

To obtain an expression for $\underset{\sim}{v}_{|\alpha}(\underset{\sim}{r}_\alpha|\underset{\sim}{r}_\alpha)$, we multiply equation 104 by $f_\alpha^{(1)}(\underset{\sim}{R}_\alpha|\underset{\sim}{r}_\alpha)$ and average over orientations of the reference molecule before proceeding further. This yields the result

$$\underset{\sim}{v}_{|\alpha}(\underset{\sim}{r}_\alpha|\underset{\sim}{r}_\alpha) = \frac{1}{6\pi\eta}\int_{r_{\beta\alpha}=\mathcal{R}_\alpha}^{r_{\beta\alpha}=+\infty} \frac{1}{r_{\beta\alpha}}\,\underset{\sim}{Y}_{|\alpha}^{(e)}(\underset{\sim}{r}_{\beta\alpha}|\underset{\sim}{r}_\alpha)\,d^3\underset{\sim}{r}_{\beta\alpha} \tag{107}$$

where $\underset{\sim}{Y}_{|\alpha}^{(e)}(\underset{\sim}{r}_{\beta\alpha}|\underset{\sim}{r}_\alpha)$ is defined in the obvious way.

IV. Electrolytic Transport of Non-electrolytes: An Approximate Theory Valid at Infinite Dilution

When a solution containing a non-electrolyte and an electrolyte is placed in a constant homogeneous electric field (a dc electric field with the constant homogeneous electric field intensity $\underset{\sim}{\underline{E}} = \underline{E}\,\underset{\sim}{\underline{k}}$) and maintained at a constant uniform temperature $\underline{T}$, the non-electrolyte usually undergoes transport. Steady state motion is set up. The velocity of the non-electrolyte molecules (hereafter referred to as species o) in the electric field is given by $\underset{\sim}{\underline{u}}_o - \underset{\sim}{\underline{u}}_\tau = (1/\underline{C}_o)\ \underset{\sim}{\underline{J}}_o^{(\mathrm{rel}\ \tau)}$ where $\underset{\sim}{\underline{J}}_o^{(\mathrm{rel}\ \tau)}$ is the diffusion current of non-electrolyte molecules defined relative to the motion of the solvent (species τ) by equation 22. This quantity is ordinarily assumed to be proportional to the electric field strength. The corresponding quantity which depends only on molecular parameters is the **mobility** of the non-electrolyte molecules

$$\underline{U}_o = (\underset{\sim}{\underline{u}}_o - \underset{\sim}{\underline{u}}_\tau)/\underset{\sim}{\underline{E}} = \underset{\sim}{\underline{J}}_o^{(\mathrm{rel}\ \tau)}/\underline{C}_o\,\underset{\sim}{\underline{E}}\,. \tag{108}$$

Alternately, we can describe the current of neutral non-electrolyte molecules produced by the transport of electrolyte ions carrying the electric current $\underset{\sim}{\underline{I}}$ in terms of the **mass** transport ratio of the non-electrolyte molecules

$$\underline{t}_o^{(\underline{m})} = \underline{m}_o \; \underset{\sim}{\underline{J}}_o^{(rel\;\tau)} / \underset{\sim}{\underline{I}} \tag{109}$$

as defined by equation 25. It is $\underline{U}_o$ and $\underline{t}_o^{(\underline{m})}$ which are ordinarily determined from experimental data.

A Model Representing the Solution Containing Non-electrolyte Molecules and Electrolyte Ions

We must now select a model which can be used to represent, schematically at least, the solution containing non-electrolyte molecules and ions of a simple electrolyte. Our classical fluid system, then, is a dilute solution containing $\underline{N}$ solute molecules or ions dissolved in a solvent (molecules of species τ) so that the total volume of the system is $\underline{V}$. There are $\underline{N}_o$ non-electrolyte molecules (species o) and $\underline{N}_a$ electrolyte ions of species a where a runs from 1 to σ.

The solvent (τ) is a structureless, viscous, incompressible dielectric continuum with dielectric constant ϵ and viscosity η.

The non-electrolyte molecules (species o) are rigid spheres, large in comparison to the solvent molecules, each having radius $\underline{b}_o$ and dielectric constant ϵ_o. Each of these spheres contains s_o discrete point charges e_i, where $i = 1, \ldots, s_o$, located at positions $\mathbf{r}_i$ relative to its center as the origin. Each such sphere has a zero net charge. The center of mass of each sphere is located at its center. Thus the non-electrolyte molecules have rotational as well as translational degrees of freedom. The kinetic entity moving in the solvent is not necessarily the bare non-electrolyte molecule of radius b_o; usually it is the non-electrolyte molecule plus a number of adjacent solvent molecules. This entity may be approximated by a sphere of radius $\mathcal{R}_o$. We shall call $\mathcal{R}_o$ the hydrodynamic radius of the non-electrolyte molecule of species o . Finally, the non-electrolyte molecules are assumed to be present at extremely low concentrations. In fact, we shall consider only those solutions in which the number density C_o of non-electrolyte molecules is so low that their mutual interactions can be neglected.

We shall be particularly interested in solutions containing non-electrolyte molecules which are dipolar ions or zwitterions. A dipolar ion is a (neutral) non-electrolyte molecule having an electric dipole moment μ_o of large magnitude. Accordingly, we represent it by a rigid sphere possessing two equal and opposite charges $e_1 > 0$ and $e_2 < 0$ situated at equal distances r_1 and r_2 from its center, with $\mathbf{r}_2 = - \mathbf{r}_1$. This charge distribution is characterized by the electric dipole moment $\boldsymbol{\mu}_o = 2\, e_1 \mathbf{r}_1$.

If we pass to the limit as r_1 approaches zero and e_1 approaches $+\infty$ in such a manner that their product remains constant, we may represent the dipolar ion by a sphere with a point or ideal dipole of moment μ_o located at its center. As mentioned before, μ_o specifies the orientation of the dipolar ion of species o (Eulerian angles Θ_o and Φ_o) and gives in addition the magnitude μ_o of its electric dipole moment.

The electrolyte ions (species α where α runs from 1 to σ) have non-zero net charges. These simple ions are assumed to be rigid spheres with spherically symmetrical charge and mass distributions, and with dielectric constants the same as that of the solvent. In other words, their multipole (dipole, quadrupole, etc.) electric moments are assumed to be zero; the center of mass of each sphere is located at its center; and possible ionic cavity effects are neglected. We let e_α be the charge of an ion of species α, and we let b_α be its radius.

The very existence of a solution of non-electrolyte molecules and electrolyte ions depends on their having finite closest distances of approach. Thus we let $a_{\alpha o}$ represent the center-to-center distance of closest approach of the ions of species α to a non-electrolyte molecule of species o.

The model which we have just described was used by

(30) Kirkwood[30] in developing an equilibrium theory of solutions containing non-electrolytes and electrolytes.

According to our picture of the solution, the various solute particles are assumed to be rigid charged spheres which interact with one another according to the laws of classical electrostatics and classical mechanics. Solutions to some of the important electrostatic problems suggested by this model are given in Appendix D. In particular, the electrostatic contribution to the mutual potential energy of a non-electrolyte molecule of species o in configuration $\mathbf{q}_o$ and of an electrolyte ion of species α in configuration $\mathbf{q}_\alpha$ is shown to have the form (equation D 10)

$$V_{o\alpha}(\mathbf{q}_o, \mathbf{q}_\alpha) = V_{o\alpha}^{LR}(\mathbf{q}_o, \mathbf{q}_\alpha) + V_{o\alpha}^{SR}(\mathbf{q}_o, \mathbf{q}_\alpha) \qquad (110)$$

where $V_{o\alpha}^{LR}(\mathbf{q}_o, \mathbf{q}_\alpha)$, which is a function of $\mathbf{R}_o$ and $\mathbf{r}_{\alpha o} = \mathbf{r}_\alpha - \mathbf{r}_o$, is the potential of the (possibly long range) Coulombic force acting between the pair (see equations D 11 and D 15); and where

$$V_{o\alpha}^{SR}(\mathbf{q}_o, \mathbf{q}_\alpha) = -\sum_{n=4}^{+\infty} d_{o\alpha,n} \left(\frac{a_{\alpha o}}{r_{\alpha o}}\right)^{n} \qquad (r_{\alpha o} \geq a_{\alpha o}) \qquad (111)$$

which is a function of the center-to-center distance $r_{\alpha o}$

between the pair, represents a short range electrostatic ion-cavity repulsive force between the ion of species α and an image distribution in the spherical cavity of low dielectric constant created by the non-electrolyte molecule in the solvent (see equations D12 and D13). As indicated, we consider short range forces to be those for which the pair potential energy is proportional to $\underline{r}_{\alpha o}^{-\underline{n}}$ where $\underline{n} \geq 4$; and we consider long range forces to be those for which the pair potential energy is proportional to $\underline{r}_{\alpha o}^{-\underline{n}}$ where $\underline{n} \leq 3$.

To account for the fact that the non-electrolyte molecules and electrolyte ions are rigid spheres which do not interpenetrate, we define

$$\underline{V}_{o\alpha}^{SR}(\underline{q}_o, \underline{q}_\alpha) = +\infty \qquad (\underline{r}_{\alpha o} < \underline{a}_{\alpha o}). \tag{112}$$

We recognize, moreover, that other short range forces may exist and may in part be responsible for the specific effects of the ions. In order to allow for these, we can allow the constant coefficients $\underline{d}_{o\alpha,\underline{n}}$ in equation 111 to be parameters which may be determined by various experimental or theoretical methods.

An Approximate Theory Valid at Infinite Dilution

Using the model just described, we now turn to the task of obtaining approximate theoretical expressions for the mobility of the non-electrolyte molecules $\underline{U}_0$ as defined by equation 108 and the mass transport ratio of the non-electrolyte molecules $\underline{t}_0^{(\underline{m})}$ as defined by equation 109. We shall consider only those solutions in which the macroscopic number densities of the non-electrolyte molecules $\underline{C}_0$ and the electrolyte ions $\underline{C}_\alpha$, where $\alpha = 1, \ldots, \sigma$, are so small that they can be neglected relative to unity. The equations, calculations, and results presented here will therefore be valid only at infinite dilution.

From equations 24 and 21, the total electric current density is

$$\underset{\sim}{I} = \underline{e} \sum_{\alpha=1}^{\sigma} \underline{C}_\alpha \, \underline{Z}_\alpha (\underset{\sim}{u}_\alpha - \underset{\sim}{u}) \quad . \tag{113}$$

Now $\underset{\sim}{u}_\alpha$, which is not a function of position since the external field is homogeneous, can be obtained from equation 75. Thus

$$\underset{\sim}{u}_\alpha(\underset{\sim}{r}_\alpha) = \underset{\sim}{u}(\underset{\sim}{r}_\alpha) + \underset{\sim}{v}_{|\alpha}(\underset{\sim}{r}_\alpha | \underset{\sim}{r}_\alpha) + \frac{\underline{D}_\alpha}{\underline{kT}} \left[\underset{\sim}{F}_\alpha^{(1)}(\underset{\sim}{r}_\alpha) + \underset{\sim}{K}_\alpha^{(1)}(\underset{\sim}{r}_\alpha) \right] . \tag{114}$$

The electrophoretic velocity $\underset{\sim}{v}_{|\alpha}(\underset{\sim}{r}_\alpha | \underset{\sim}{r}_\alpha)$ is given by equation 107; $\underset{\sim}{F}_\alpha^{(1)}(\underset{\sim}{r}_\alpha)$, which represents the force due to the external field as well as the relaxation force, is given by equation 29; and the kinetic force $\underset{\sim}{K}_\alpha^{(1)}(\underset{\sim}{r}_\alpha)$ is given by equation 103. For an electrolyte ion of species α, the force on the ion due to the external field $\underset{\sim}{E}$ is $\underset{\sim}{F}_\alpha^{(E)}(\underset{\sim}{q}_\alpha) = e_\alpha \underset{\sim}{E}$ as given by equation D 27. Since we are neglecting terms of order C_β, where $\beta = 1, \ldots, \sigma$, relative to unity; we have

$$\underset{\sim}{F}_\alpha^{(1)}(\underset{\sim}{r}_\alpha) = e Z_\alpha \underset{\sim}{E} \tag{115}$$

and

$$\underset{\sim}{u}_\alpha(\underset{\sim}{r}_\alpha) - \underset{\sim}{u}(\underset{\sim}{r}_\alpha) = \frac{D_\alpha Z_\alpha e}{kT} \underset{\sim}{E} \tag{116}$$

so that the total electric current density is simply

$$\underset{\sim}{I} = (\underset{\sim}{E}/kT) \sum_{\alpha=1}^{\sigma} D_\alpha C_\alpha e_\alpha^2 . \tag{117}$$

From equations 23, 21, and 75, the diffusion current for the non-electrolyte, which is not a function of position since the external field is homogeneous, is given by

$$\underline{J}_o^{(rel\,\tau)}(\underline{r}_o) = \underline{c}_o(\underline{u}_o(\underline{r}_o) - \underline{u}(\underline{r}_o)) = \tag{118}$$

$$= \underline{c}_o \left[\underline{v}_{|o}(\underline{r}_o | \underline{r}_o) + \frac{\underline{D}_o}{\underline{kT}} \underline{F}_o^{(1)}(\underline{r}_o) + \frac{\underline{D}_o}{\underline{kT}} \underline{K}_o^{(1)}(\underline{r}_o) \right].$$

The external force $\underline{F}_o^{(\underline{E})}(\underline{q}_o)$ on a neutral non-electrolyte molecule is zero. Therefore, the relaxation force as given by equations 29, 50, 51, and 53 is

$$\underline{F}_o^{(1)}(\underline{r}_o) = \frac{1}{64\,\pi^4} \sum_{\alpha=1}^{\sigma} \underline{c}_\alpha \iint \underline{F}_{\alpha o}(\underline{q}_\alpha, \underline{q}_o) \; \underline{g}_{o\alpha}^{(2)}(\underline{q}_o ; \underline{q}_\alpha) \; \underline{d}^3\underline{R}_o \; \underline{d}^6\underline{q}_\alpha . \tag{119}$$

The kinetic force, as given by equation 103, is

$$\underline{K}_o^{(1)}(\underline{r}_o) = \tag{120}$$

$$= -\underline{kT} \sum_{\alpha=1}^{\sigma} \underline{c}_\alpha \iint \underline{g}_{o\alpha}^{(2)}(\underline{r}_o ; \underline{r}_o + \mathcal{R}_o \, \underline{1}\underline{r}_{\alpha o}) \; \underline{1}\underline{r}_{\alpha o} \, \mathcal{R}_o^2 \sin \theta_{\alpha o} \; \underline{d}\theta_{\alpha o} \; \underline{d}\varphi_{\alpha o} .$$

From equation 107, the electrophoretic velocity is

$$\underset{\sim}{v}|_{0}(\underset{\sim}{r}_0|\underset{\sim}{r}_0) = \frac{1}{6\pi\eta}\underset{\sim}{E}\sum_{\alpha=1}^{\sigma} C_\alpha e_\alpha \int_{r_{\alpha o}=\mathcal{R}_o}^{r_{\alpha o}=+\infty} \frac{1}{r_{\alpha o}}\left[g_{o\alpha}^{(2,0)}(r_{\alpha o}) - 1\right] d^3\underset{\sim}{r}_{\alpha o} \tag{121}$$

since we are neglecting terms of order E^2 relative to terms of order E. In order to proceed further we must obtain an approximate expression for the $o\alpha$-pair correlation function $g_{o\alpha}^{(2)}(\underset{\sim}{q}_o ; \underset{\sim}{q}_\alpha)$, where o represents a non-electrolyte molecule and α represents an electrolyte ion. Using equation 44, we write $g_{o\alpha}^{(2)}(\underset{\sim}{q}_o ; \underset{\sim}{q}_\alpha)$ as the sum of an equilibrium term $g_{o\alpha}^{(2,0)}(\underset{\sim}{q}_o ; \underset{\sim}{q}_\alpha)$ of order zero in the electric field E plus a first order perturbation $g_{o\alpha}^{(2,1)}(\underset{\sim}{q}_o ; \underset{\sim}{q}_\alpha)$ proportional to E.

When terms of order $1/N_o$ and $1/N_\alpha$, where $\alpha = 1,\ldots,\sigma$, and terms of order C_o and C_α, where $\alpha = 1,\ldots,\sigma$, are neglected relative to unity; the equilibrium $o\alpha$-pair correlation function as given by equation 39 is

$$g_{o\alpha}^{(2,0)}(\underset{\sim}{q}_o ; \underset{\sim}{q}_\alpha) = \exp\left[-\frac{1}{kT} V_{o\alpha}(\underset{\sim}{q}_o , \underset{\sim}{q}_\alpha)\right] .$$

Here $V_{o\alpha}(\underset{\sim}{q}_o , \underset{\sim}{q}_\alpha)$ is the mutual potential energy of the pair as given by equation 110. In solvents of relatively high dielectric constant at ordinary temperatures, a satisfactory approximation[31] to $g_{o\alpha}^{(2,0)}(\underset{\sim}{q}_o ; \underset{\sim}{q}_\alpha)$ may be obtained by expanding the above exponential expression and retaining terms up to order $(1/kT)^2$ only. This yields:

$$g_{o\alpha}^{(2,0)}(q_o; q_\alpha) = 1 - \frac{1}{kT} V_{o\alpha}(q_o; q_\alpha) + \frac{1}{2}\left[\frac{1}{kT} V_{o\alpha}(q_o; q_\alpha)\right]^2 \tag{122}$$

provided $r_{\alpha o} \geq a_{\alpha o}$.

An approximate expression for $g_{o\alpha}^{(2,1)}(q_o; q_\alpha)$ can be obtained by integrating the equation of continuity in $o\alpha$-pair configuration space, i.e., by solving equation 91. When terms of order C_o and C_α and terms of order $1/N_o$ and $1/N_\alpha$, where $\alpha = 1, \ldots, \sigma$, are neglected relative to unity; this monstrous equation degenerates into

$$(D_o + D_\alpha) \nabla_{r_{\alpha o}}^2 g_{o\alpha}^{(2,1)} + D_o^{(rot)} \nabla_{R_o}^2 g_{o\alpha}^{(2,1)} + \tag{123}$$

$$\left[\frac{D_o}{kT} F_{\alpha o} - \frac{D_\alpha}{kT} F_{o\alpha}\right] \cdot \nabla_{r_{\alpha o}} g_{o\alpha}^{(2,1)} +$$

$$- \frac{D_o^{(rot)}}{kT} T_{\alpha o} \cdot \nabla_{R_o} g_{o\alpha}^{(2,1)} +$$

$$\left[\frac{D_o}{kT} \nabla_{r_{\alpha o}} \cdot F_{\alpha o} - \frac{D_\alpha}{kT} \nabla_{r_{\alpha o}} \cdot F_{o\alpha} - \frac{D_o^{(rot)}}{kT} \nabla_{R_o} \cdot T_{\alpha o}\right] g_{o\alpha}^{(2,1)}$$

$$= \frac{D_\alpha}{kT}\left[\underset{\approx}{F}_\alpha^{(E)} + \underset{\approx}{F}_{o\alpha}^{(E)}\right] \cdot \nabla_{\underset{\sim}{r}_{\alpha o}} g_{o\alpha}^{(2,0)} +$$

$$+ \frac{D_o^{(rot)}}{kT} \underset{\approx}{T}_o^{(E)} \cdot \nabla_{\underset{\sim}{R}_o} g_{o\alpha}^{(2,0)} +$$

$$+ \frac{D_o^{(rot)}}{kT}\left[\nabla_{\underset{\sim}{R}_o} \cdot \underset{\approx}{T}_o^{(E)}\right] g_{o\alpha}^{(2,0)} .$$

A number of terms appearing in equation 91 have vanished because of the properties of the neutral non-electrolyte molecules of species o , because of the properties of the electrolyte ions of species α , and because the external field is a homogeneous electric field $\underset{\sim}{E}$.

The hydrodynamic boundary condition on $g_{o\alpha}^{(2,1)}(\underset{\sim}{q}_o ; \underset{\sim}{q}_\alpha)$ arising from the requirement that rigid spheres do not interpenetrate, as expressed by equation 92, degenerates into

$$\Bigg[(D_o + D_\alpha) \nabla_{\underset{\sim}{r}_{\alpha o}} g_{o\alpha}^{(2,1)} + \left(\frac{D_o}{kT} \underset{\sim}{F}_{\alpha o} - \frac{D_\alpha}{kT} \underset{\sim}{F}_{o\alpha}\right) g_{o\alpha}^{(2,1)} + - \frac{D_\alpha}{kT}\left(\underset{\approx}{F}_\alpha^{(E)} + \underset{\approx}{F}_{o\alpha}^{(E)}\right) g_{o\alpha}^{(2,0)}\Bigg]_{r_{\alpha o} = a_{\alpha o}} \cdot \underset{\sim}{1}_{r_{\alpha o}} = 0. \tag{124}$$

The requirement that there be no correlation at infinite separation of the two spheres in a fluid, equation 46, becomes

$$\lim_{\underline{r}_{\alpha o} \to +\infty} \underline{g}_{o\alpha}^{(2,1)}(\underset{\sim}{q}_o ; \underset{\sim}{q}_\alpha) = \frac{\underline{c}_o^{(1,1)}(\underset{\sim}{q}_o)}{\underline{c}_o^{(1,0)}(\underset{\sim}{q}_o)} \quad . \tag{125}$$

In order to evaluate $\underline{c}_o^{(1,1)}(\underset{\sim}{q}_o)$, we must integrate the equation of continuity in o-configuration space, i.e., we must solve equation 88. In the present context this equation reduces to

$$\nabla_{\underset{\sim}{R}_o}^2 \underline{c}_o^{(1,1)}(\underset{\sim}{q}_o) = \frac{1}{\underline{kT}} \underline{c}_o^{(1,0)}(\underset{\sim}{q}_o) \nabla_{\underset{\sim}{R}_o} \cdot \underset{\sim}{T}_o^{(\underline{E})}(\underset{\sim}{q}_o) \tag{126}$$

Since $\underset{\sim}{T}_o^{(\underline{E})}(\underset{\sim}{q}_o) = - \nabla_{\underset{\sim}{R}_o} \underline{V}_o^{(\underline{E})}(\underset{\sim}{q}_o)$, the obvious solution is

$$\underline{c}_o^{(1,1)}(\underset{\sim}{q}_o) = - \frac{1}{\underline{kT}} \underline{c}_o^{(1,0)}(\underset{\sim}{q}_o) \underline{V}_o^{(\underline{E})}(\underset{\sim}{q}_o). \tag{127}$$

Our boundary condition, equation 125, becomes

$$\lim_{\underline{r}_{\alpha o} \to +\infty} \underline{g}_{o\alpha}^{(2,1)}(\underset{\sim}{q}_o ; \underset{\sim}{q}_\alpha) = - \frac{1}{\underline{kT}} \underline{V}_o^{(\underline{E})}(\underset{\sim}{q}_o). \tag{128}$$

Once equation 123 is solved and an expression for $\underline{g}_{o\alpha}^{(2)}(\underset{\sim}{q}_o;\underset{\sim}{q}_\alpha)$ is obtained; the mobility $\underline{U}_o$ may be calculated from equations 118 and 108, and the mass transport ratio $\underline{t}_o^{(\underline{m})}$ may be calculated from equations 118, 117, and 109. In the next section of this paper, we shall calculate $\underline{U}_o$ and $\underline{t}_o^{(\underline{m})}$ for an ideal dipolar non-electrolyte.

Electrolytic Transport of an Ideal Dipolar Non-electrolyte

We now assume that the non-electrolyte molecule is an ideal dipolar ion with a point or ideal dipole of moment $\underset{\sim}{\mu}_o$ located at its center.

For this case, the pair potential energy $\underline{V}_{o\alpha}(\underset{\sim}{q}_o, \underset{\sim}{q}_\alpha)$ is given by equations 110, D 15, 111 and 112. From equation 122 we obtain as an approximation to the equilibrium $o\alpha$-pair correlation function the expression

$$\underline{g}_{o\alpha}^{(2,0)}(\underset{\sim}{q}_o;\underset{\sim}{q}_\alpha) = 1 + \sum_{\underline{n}=4}^{+\infty} \frac{\underline{d}_{o\alpha,\underline{n}}}{\underline{kT}} \left(\frac{\underline{a}_{\alpha o}}{\underline{r}_{\alpha o}}\right)^{\underline{n}} + \tag{129}$$

$$+ \frac{1}{2} \sum_{\underline{n},\underline{l}=4}^{+\infty} \frac{\underline{d}_{o\alpha,\underline{n}}}{\underline{kT}} \frac{\underline{d}_{o\alpha,\underline{l}}}{\underline{kT}} \left(\frac{\underline{a}_{\alpha o}}{\underline{r}_{\alpha o}}\right)^{\underline{n}+\underline{l}} +$$

$$- \frac{3 e_\alpha}{(2\epsilon + \epsilon_o)(kT)} \frac{1}{r_{\alpha o}^2} \underset{\sim}{\mu}_o \cdot \underset{\sim}{1} r_{\alpha o} +$$

$$- \frac{3 e_\alpha}{(2\epsilon + \epsilon_o)(kT)} \underset{\sim}{\mu}_o \cdot \underset{\sim}{1} r_{\alpha o} \sum_{n=4}^{+\infty} \frac{d_{o\alpha,n}}{kT} \frac{a_{\alpha o}^n}{r_{\alpha o}^{n+2}} +$$

$$+ \frac{9 e_\alpha^2}{2(2\epsilon + \epsilon_o)^2 (kT)^2} \frac{1}{r_{\alpha o}^4} \underset{\sim}{\mu}_o \cdot \underset{\sim}{1} r_{\alpha o} \underset{\sim}{\mu}_o \cdot \underset{\sim}{1} r_{\alpha o}$$

provided $r_{\alpha o} \geq a_{\alpha o}$.

Upon averaging over orientations of the non-electrolyte molecule and of the electrolyte ion — see equation 32 and Appendix C — we obtain the equilibrium $o\alpha$-pair position correlation function

$$g_{o\alpha}^{(2,0)}(r_{\alpha o}) = 1 + \sum_{n=4}^{+\infty} \frac{d_{o\alpha,n}}{kT} \left(\frac{a_{\alpha o}}{r_{\alpha o}}\right)^n + \tag{130}$$

$$+ \frac{1}{2} \sum_{n,l=4}^{+\infty} \frac{d_{o\alpha,n}}{kT} \frac{d_{o\alpha,l}}{kT} \left(\frac{a_{\alpha o}}{r_{\alpha o}}\right)^{n+l} +$$

$$+ \quad \frac{3 e_\alpha^2 \; \mu_o^2}{2(2\epsilon + \epsilon_o)^2 (kT)^2} \quad \frac{1}{r_{\alpha o}^4} \quad .$$

Next, we must obtain an approximate expression for the non-equilibrium perturbation $g_{o\alpha}^{(2,1)}(q_o; q_\alpha)$ to the $o\alpha$-pair correlation function. Since the non-electrolyte molecule is an ideal dipolar ion, $F_{\alpha o}(q_\alpha, q_o) = - F_{o\alpha}(q_o, q_\alpha)$ is given by equation D 16; $T_{\alpha o}(q_\alpha, q_o)$ is given by equation D 17; $V_o^{(E)}(q_o)$ is given by equation D 23; $T_o^{(E)}(q_o)$ is given by equation D 25; and $F_{o\alpha}^{(E)}(q_o, q_\alpha; \xi_o = 0)$ is given by equation D 30. Of course $F_\alpha^{(E)}(q_\alpha)$ is given by equation D 27. The diffusion constants D_o, D_α, and $D_o^{(rot)}$ may be determined experimentally or theoretically — possibly using equations 73 and 81. Thus the only unknown quantity in equations 123, 124, and 128 is $g_{o\alpha}^{(2,1)}(q_o; q_\alpha)$. Upon retaining only quadratic terms in $(1/kT)$ — since only quadratic terms in $(1/kT)$ were retained in the approximate expression for $g_{o\alpha}^{(2,0)}(q_o; q_\alpha)$ given by equation 129 — the equation of continuity in the pair configuration space of an ideal dipolar non-electrolyte molecule of species o and an electrolyte ion of species α (equation 123) becomes

(131)

$$\nabla_{r_{\alpha o}}^2 \; g_{o\alpha}^{(2,1)} \quad + \quad \frac{1}{2} \lambda_\alpha^2 \; \nabla_{R_o}^2 \; g_{o\alpha}^{(2,1)} \quad +$$

$$\sum_{n=4}^{+\infty} n \frac{d_{o\alpha,n}}{kT} \frac{(a_{\alpha o})^{n}}{(r_{\alpha o})^{n+1}} \; 1_{r_{\alpha o}} \cdot \nabla_{r_{\alpha o}} \; g_{o\alpha}^{(2,1)} \quad +$$

$$\frac{3\,e_{\alpha}}{(2\epsilon + \epsilon_{o})\,kT} \frac{1}{r_{\alpha o}^{3}} \left[\mu_{o} - 3\left(\mu_{o} \cdot 1_{r_{\alpha o}}\right) 1_{r_{\alpha o}} \right] \cdot \nabla_{r_{\alpha o}} \; g_{o\alpha}^{(2,1)} \; +$$

$$\frac{3\,e_{\alpha}\,\lambda_{\alpha}^{2}}{2(2\epsilon + \epsilon_{o})\,kT} \frac{1}{r_{\alpha o}^{2}} \left[\nabla_{R_{o}} \left| \mu_{o} \cdot 1_{r_{\alpha o}} \right| \right] \cdot \nabla_{R_{o}} \quad g_{o\alpha}^{(2,1)} \quad +$$

$$\sum_{n=4}^{+\infty} n(1-n) \frac{d_{o\alpha,n}}{kT} \frac{(a_{\alpha o})^{n}}{(r_{\alpha o})^{n+2}} \; g_{o\alpha}^{(2,1)} \quad +$$

$$- \; \frac{3\,e_{\alpha}\;\lambda_{\alpha}^{2}}{(2\epsilon + \epsilon_{o})\,kT} \frac{1}{r_{\alpha o}^{2}} \; \mu_{o} \cdot 1_{r_{\alpha o}} \; g_{o\alpha}^{(2,1)}$$

$$= \frac{D_{\alpha}\;e_{\alpha}}{(D_{o} + D_{\alpha})\,kT} \, E \cdot \nabla_{r_{\alpha o}} \left[\sum_{n=4}^{+\infty} \frac{d_{o\alpha,n}}{kT} \left(\frac{a_{\alpha o}}{r_{\alpha o}} \right)^{n} \right] \quad +$$

$$- \; \frac{D_{\alpha}\,e_{\alpha}\;B_{o}^{3}}{(D_{o} + D_{\alpha})\,kT} \, E \cdot \nabla_{r_{\alpha o}} \left[\sum_{n=4}^{+\infty} \frac{n}{n+3} \; \frac{d_{o\alpha,n}}{kT} \; \frac{(a_{\alpha o})^{n}}{(r_{\alpha o})^{n+3}} \right] +$$

$$- \frac{3\lambda_{\alpha}^{2}\epsilon}{(2\epsilon + \epsilon_{o})\,\underline{kT}} \left[1 + \sum_{\underline{n}=4}^{+\infty} \frac{d_{o\alpha,\underline{n}}}{\underline{kT}} \left(\frac{a_{\alpha o}}{r_{\alpha o}}\right)^{\underline{n}} \right] \underset{\approx}{E} \cdot \underset{\sim}{\mu}_{o} +$$

$$- \frac{9\,\underline{D}_{\alpha}\,e_{\alpha}^{2}}{4(\underline{D}_{o} + \underline{D}_{\alpha})(2\epsilon + \epsilon_{o})(\underline{kT})^{2}} \frac{B_{o}^{3}}{r_{\alpha o}^{6}} \underset{\approx}{E} \cdot \underset{\sim}{\mu}_{o} +$$

$$+ \frac{3\,\underline{D}_{\alpha}\,e_{\alpha}^{2}}{(\underline{D}_{o} + \underline{D}_{\alpha})(2\epsilon + \epsilon_{o})(kT)^{2}} \underset{\approx}{E}\underset{\sim}{\mu}_{o} : \nabla_{\underset{\sim}{r}_{\alpha o}} \nabla_{\underset{\sim}{r}_{\alpha o}} \left[\frac{1}{r_{\alpha o}} - \frac{B_{o}^{3}}{16\, r_{\alpha o}^{4}} \right] +$$

$$+ \frac{9\epsilon\lambda_{\alpha}^{2}\, e_{\alpha}}{2(2\epsilon + \epsilon_{o})^{2}(\underline{kT})^{2}} \left[\mu_{o}^{2}\, \underset{\approx}{E} \cdot \nabla_{\underset{\sim}{r}_{\alpha o}} \frac{1}{r_{\alpha o}} - 3\, \underset{\approx}{E} \cdot \underset{\sim}{\mu}_{o}\, \underset{\sim}{\mu}_{o} \cdot \nabla_{\underset{\sim}{r}_{\alpha o}} \frac{1}{r_{\alpha o}} \right]$$

where, by definition

$$\lambda_{\alpha}^{2} = \frac{2\,\underline{D}_{o}^{(rot)}}{\underline{D}_{o} + \underline{D}_{\alpha}} \tag{132}$$

and

$$\underline{B}_{o}^{3} = \frac{2(\epsilon - \epsilon_{o})}{2\epsilon + \epsilon_{o}}\, \underline{b}_{o}^{3} \; . \tag{133}$$

With the same approximations, the hydrodynamic boundary condition on $g_{o\alpha}^{(2,1)}(q_o; q_\alpha)$, equation 124, becomes

$$\Bigg[\frac{\partial g_{o\alpha}^{(2,1)}}{\partial r_{\alpha o}} + \sum_{n=4}^{+\infty} n \frac{d_{o\alpha,n}}{kT} \frac{(a_{\alpha o})^n}{(r_{\alpha o})^{n+1}} g_{o\alpha}^{(2,1)} + \tag{134}$$

$$- \frac{6 e_\alpha}{(2\epsilon + \epsilon_o) kT} \frac{1}{r_{\alpha o}^3} \mu_o \cdot 1 r_{\alpha o} \, g_{o\alpha}^{(2,1)} +$$

$$- \frac{D_\alpha e_\alpha}{(D_o + D_\alpha) kT} \left(1 - \frac{B_o^3}{r_{\alpha o}^3}\right) \left(1 + \sum_{n=4}^{+\infty} \frac{d_{o\alpha,n}}{kT} \left(\frac{a_{\alpha o}}{r_{\alpha o}}\right)^n\right) E \cdot 1 r_{\alpha o} +$$

$$+ \frac{3 D_\alpha e_\alpha^2}{(D_o + D_\alpha)(2\epsilon + \epsilon_o)(kT)^2} \left(1 - \frac{B_o^3}{r_{\alpha o}^3}\right) \frac{1}{r_{\alpha o}^2} E \mu_o : 1 r_{\alpha o} 1 r_{\alpha o} \Bigg]_{r_{\alpha o} = a_{\alpha o}}$$

$$= 0 .$$

Equation 127 becomes

$$c_o^{(1,1)}(q_o) = \frac{3\epsilon}{2\epsilon + \epsilon_o} \frac{1}{kT} c_o^{(1,0)}(q_o) \, E \cdot \mu_o , \tag{135}$$

and the second boundary condition on $\underline{g}_{0\alpha}^{(2,1)}(\underset{\sim}{q}_0 ; \underset{\sim}{q}_\alpha)$, the requirement of no correlation at infinite separation of the two spheres in the fluid, equation 128, becomes

$$\lim_{\underline{r}_{\alpha 0} \to +\infty} \underline{g}_{0\alpha}^{(2,1)}(\underset{\sim}{q}_0 ; \underset{\sim}{q}_\alpha) = \tag{136}$$

$$= \frac{3\epsilon}{2\epsilon + \epsilon_0} \frac{1}{\underline{kT}} \underset{\sim}{E} \cdot \underset{\sim}{\mu}_0 .$$

Upon expanding $\underline{g}_{0\alpha}^{(2,1)}(\underset{\sim}{q}_0 ; \underset{\sim}{q}_\alpha)$ in powers of $(1/\underline{kT})$, we observe that the first two terms on the left-hand side of equation 131 are of lower order in $(1/\underline{kT})$ than are the other terms on that side. This suggests that an iterative procedure be used in solving this equation. As a first step we shall neglect quadratic terms in $(1/\underline{kT})$ relative to terms linear in $(1/\underline{kT})$. By solving the resulting equation of continuity we can obtain an expression for $\underline{g}_{0\alpha}^{(2,1)}(\underset{\sim}{q}_0 ; \underset{\sim}{q}_\alpha)$ which is valid up to terms linear in $(1/\underline{kT})$.

When only linear terms in $(1/\underline{kT})$ are retained, equation 131 becomes simply

$$\nabla_{\underset{\sim}{r}_{\alpha 0}}^2 \underline{g}_{0\alpha}^{(2,1)} + \frac{1}{2} \lambda_\alpha^2 \nabla_{\underset{\sim}{R}_0}^2 \underline{g}_{0\alpha}^{(2,1)} \tag{137}$$

$$= \quad - \frac{3 \lambda_\alpha^2 \epsilon}{(2\epsilon + \epsilon_o)\, \underline{kT}} \quad \underset{\sim}{\underline{E}} \cdot \underset{\sim}{\mu}_o \, ,$$

and the boundary conditions on $\underline{g}_{o\alpha}^{(2,1)}(\underset{\sim}{\underline{q}}_o ; \underset{\sim}{\underline{q}}_\alpha)$ are equation 136 and, from equation 134,

$$\left. \frac{\partial\, \underline{g}_{o\alpha}^{(2,1)}}{\partial\, \underline{r}_{\alpha o}} \right]_{\underline{r}_{\alpha o} = \underline{a}_{\alpha o}} = \tag{138}$$

$$= \quad \frac{\underline{D}_\alpha \; \underline{e}_\alpha}{(\underline{D}_o + \underline{D}_\alpha)\; \underline{kT}} \left(1 - \frac{\underline{B}_o^3}{\underline{a}_{\alpha o}^3} \right) \underset{\sim}{\underline{E}} \cdot \underset{\sim}{\underline{1}}_{\underline{r}_{\alpha o}} \, .$$

In order to solve equation 137 we assume that a solution exists, expressible in the form

$$\underline{g}_{o\alpha}^{(2,1)} = \underline{f}^{(0)} + \underset{\sim}{\underline{E}} \cdot \nabla_{\underset{\sim}{\underline{r}}_{\alpha o}} \underline{f}^{(1)} + \underline{g}^{(1)} \underset{\sim}{\underline{E}} \cdot \underset{\sim}{\mu}_o \tag{139}$$

where $\underline{f}^{(0)}$, $\underline{f}^{(1)}$, and $\underline{g}^{(1)}$ are functions of $\underline{r}_{\alpha o}$ only. The term $\underline{f}^{(0)}$ is a surface harmonic of degree zero; the term $\dfrac{d\,\underline{f}^{(1)}}{d\,\underline{r}_{\alpha o}} \underset{\sim}{\underline{E}} \cdot \underset{\sim}{\underline{1}}_{\underline{r}_{\alpha o}}$ is a surface harmonic of degree one on the sphere $r_{\alpha o}$ = constant; and the term $g^{(1)} \mu_o \underset{\sim}{E} \cdot \underset{\sim}{1}_{R_o}$ is a surface harmonic of degree one on the sphere R_o = constant. Some useful properties of these surface harmonics are discussed briefly in Appendix A. If we now substitute equation 139 into equation 137, expand in surface harmonics, equate the coefficients of the linearly independent surface harmonics to zero, and neglect certain inconsequential constants of integration; we obtain three second order differential equations

$$\frac{d^2 f^{(0)}}{d\, r_{\alpha o}^2} + \frac{2}{r_{\alpha o}} \frac{d f^{(0)}}{d r_{\alpha o}} = 0 \tag{140}$$

$$\frac{d^2 f^{(1)}}{d\, r_{\alpha o}^2} + \frac{2}{r_{\alpha o}} \frac{d f^{(1)}}{d r_{\alpha o}} = 0$$

$$\frac{d^2 g^{(1)}}{d\, r_{\alpha o}^2} + \frac{2}{r_{\alpha o}} \frac{d g^{(1)}}{d r_{\alpha o}} - \lambda_\alpha^2 g^{(1)} =$$

$$= - \frac{3\epsilon}{2\epsilon + \epsilon_o} \frac{1}{kT} \lambda_\alpha^2 .$$

Solutions of these three differential equations are

$$f^{(0)} = N^{(0)} \frac{1}{r_{\alpha o}} + M^{(0)} \tag{141}$$

$$f^{(1)} = N^{(1)} \frac{1}{r_{\alpha o}} + M^{(1)}$$

$$g^{(1)} = Q^{(1)} \frac{1}{\lambda_\alpha r_{\alpha o}} \exp\left[-\lambda_\alpha r_{\alpha o}\right] +$$

$$+ P^{(1)} \frac{1}{\lambda_\alpha r_{\alpha o}} \exp\left[\lambda_\alpha r_{\alpha o}\right] + \frac{3\epsilon}{2\epsilon + \epsilon_o} \frac{1}{kT} .$$

From equation 136 both $M^{(0)}$ and $P^{(1)}$ must be zero; from equation 138 both $N^{(0)}$ and $Q^{(1)}$ must be zero, and

$$N^{(1)} = \frac{1}{2} \frac{D_\alpha e_\alpha a_{\alpha o}^3}{(D_o + D_\alpha) kT} \left[1 - \frac{B_o^3}{a_{\alpha o}^3}\right] . \tag{142}$$

Therefore, through terms linear in $(1/kT)$, the non-equilibrium perturbation to the $o\alpha$-pair correlation function is

$$g_{o\alpha}^{(2,1)}(\underset{\sim}{q}_o ; \underset{\sim}{q}_\alpha) = -\frac{1}{2} \frac{D_\alpha e_\alpha}{(D_o + D_\alpha) kT} \left[a_{\alpha o}^3 - B_o^3\right] \frac{1}{r_{\alpha o}^2} \underset{\sim}{E} \cdot \underset{\sim}{1}_{r_{\alpha o}} \tag{143}$$

$$+ \quad \frac{3\epsilon}{2\epsilon + \epsilon_o} \quad \frac{1}{\underline{kT}} \quad \underset{\sim}{\underline{E}} \cdot \underset{\sim}{\mu}_o \ .$$

If we now replace $\underline{g}_{o\alpha}^{(2,1)}$ in the previously neglected terms of equation 131 by 143, we obtain an approximate equation of continuity valid through terms quadratic in $(1/\underline{kT})$. The result is

$$\nabla_{\underset{\sim}{\underline{r}}\alpha o}^{2} \ \underline{g}_{o\alpha}^{(2,1)} + \frac{1}{2} \lambda_\alpha^2 \ \nabla_{\underset{\sim}{\underline{R}}_o}^{2} \ \underline{g}_{o\alpha}^{(2,1)} = \tag{144}$$

$$\frac{\underline{D}_\alpha \underline{e}_\alpha}{(\underline{D}_o + \underline{D}_\alpha)\,\underline{kT}} \ \underset{\sim}{\underline{E}} \cdot \nabla_{\underset{\sim}{\underline{r}}\alpha o} \left(\sum_{\underline{n}=4}^{+\infty} \frac{\underline{d}_{o\alpha,\underline{n}}}{\underline{kT}} \left(\frac{\underline{a}_{\alpha o}}{\underline{r}_{\alpha o}}\right)^{\underline{n}} \left[1 + \right.\right.$$

$$\left.\left. + \ \frac{\underline{n}(\underline{n}+1)}{2(\underline{n}+3)} \left(\frac{\underline{a}_{\alpha o}}{\underline{r}_{\alpha o}}\right)^3 - \frac{\underline{n}}{2} \left(\frac{\underline{B}_o}{\underline{r}_{\alpha o}}\right)^3 \right] \right) +$$

$$- \ \frac{3\epsilon \lambda_\alpha^2}{(2\epsilon + \epsilon_o)\,\underline{kT}} \left(1 + \sum_{\underline{n}=4}^{+\infty} \frac{\underline{d}_{o\alpha,\underline{n}}}{\underline{kT}} \left(\frac{\underline{a}_{\alpha o}}{\underline{r}_{\alpha o}}\right)^{\underline{n}} \left[1 - \frac{\underline{n}(\underline{n}-1)}{\lambda_\alpha^2 \ \underline{r}_{\alpha o}^2} \right] \right) \underset{\sim}{\underline{E}} \cdot \underset{\sim}{\mu}_o$$

$$+ \ \frac{3\,\underline{D}_\alpha \underline{e}_\alpha^2}{(\underline{D}_o + \underline{D}_\alpha)(2\epsilon + \epsilon_o)(\underline{kT})^2} \left(\frac{1}{\underline{r}_{\alpha o}^6} \left[\underline{a}_{\alpha o}^3 - 2\,\underline{B}_o^3 \right] + \right.$$

$$- \frac{\lambda_{\alpha}^{2}}{6\,\underline{r}_{\alpha o}^{4}} \left[\underline{a}_{\alpha o}^{3} - \underline{B}_{o}^{3}\right]\Bigg) \quad \underset{\sim}{\underline{E}} \cdot \underset{\sim}{\mu}_{o} \quad +$$

$$+ \frac{3\,\underline{D}_{\alpha}\,\underline{e}_{\alpha}^{2}}{(\underline{D}_{o} + \underline{D}_{\alpha})(2\epsilon + \epsilon_{o})\,(\underline{kT})^{2}} \left(\frac{3}{\underline{r}_{\alpha o}^{3}} + \frac{3}{2\,\underline{r}_{\alpha o}^{6}} \left[\underline{a}_{\alpha o}^{3} - 2\,\underline{B}_{o}^{3}\right] + \right.$$

$$\left. - \frac{\lambda_{\alpha}^{2}}{2\,\underline{r}_{\alpha o}^{4}} \left[\underline{a}_{\alpha o}^{3} - \underline{B}_{o}^{3}\right] \right) \left(\underset{\sim}{\underline{E}} \cdot \underset{\sim}{1}\underline{r}_{\alpha o}\,\underset{\sim}{\mu}_{o} \cdot \underset{\sim}{1}\underline{r}_{\alpha o} - \frac{1}{3}\underset{\sim}{\underline{E}} \cdot \underset{\sim}{\mu}_{o} \right) +$$

$$- \frac{27\,\underline{e}_{\alpha}\,\epsilon\,\lambda_{\alpha}^{2}}{(2\epsilon + \epsilon_{o})^{2}\,(\underline{kT})^{2}} \left(\underset{\sim}{\underline{E}} \cdot \underset{\sim}{\mu}_{o} \;\; \underset{\sim}{\mu}_{o} \cdot \nabla_{\underset{\sim}{\underline{r}}_{\alpha o}} \frac{1}{\underline{r}_{\alpha o}} - \frac{1}{3}\,\mu_{o}^{2}\;\underset{\sim}{\underline{E}} \cdot \nabla_{\underset{\sim}{\underline{r}}_{\alpha o}} \frac{1}{\underline{r}_{\alpha o}} \right) .$$

From equations 134 and 143 the hydrodynamic boundary condition on $\underline{g}_{o\alpha}^{(2,1)}(\underset{\sim}{\underline{q}}_{o};\underset{\sim}{\underline{q}}_{\alpha})$ valid through terms of order $(1/\underline{kT})^{2}$ is

$$\left. \frac{\partial\,\underline{g}_{o\alpha}^{(2,1)}}{\partial\,\underline{r}_{\alpha o}} \right]_{\underline{r}_{\alpha o} = \underline{a}_{\alpha o}} = \tag{145}$$

$$= \frac{\underline{D}_{\alpha}\,\underline{e}_{\alpha}}{(\underline{D}_{o} + \underline{D}_{\alpha})\,\underline{kT}} \left[1 - \left(\frac{\underline{B}_{o}}{\underline{a}_{\alpha o}}\right)^{3}\right] \left[1 + \sum_{\underline{n}=4}^{+\infty} \frac{\underline{n}+2}{2}\,\frac{\underline{d}_{o\alpha,\underline{n}}}{\underline{kT}}\right] \underset{\sim}{\underline{E}} \cdot \underset{\sim}{1}\underline{r}_{\alpha o} \quad +$$

$$+ \frac{6\epsilon\,\underline{e}_{\alpha}\,\mu_{o}^{2}}{(2\epsilon + \epsilon_{o})^{2}\,(\underline{kT})^{2}\,\underline{a}_{\alpha o}^{3}} \;\underset{\sim}{\underline{E}} \cdot \underset{\sim}{1}\underline{r}_{\alpha o} \quad +$$

$$- \frac{3\epsilon}{2\epsilon + \epsilon_o} \frac{1}{\underline{kT}} \sum_{\underline{n}=4}^{+\infty} \frac{\underline{d}_{o\alpha,\underline{n}}}{\underline{kT}} \frac{\underline{n}}{\underline{a}_{\alpha o}} \underset{\sim}{\underline{E}} \cdot \underset{\sim}{\mu}_o +$$

$$- \frac{2\,\underline{D}_\alpha\,\underline{e}_\alpha^2}{(\underline{D}_o + \underline{D}_\alpha)(2\epsilon + \epsilon_o)(\underline{kT})^2} \left[1 - \left(\frac{\underline{B}_o}{\underline{a}_{\alpha o}}\right)^3\right] \frac{1}{\underline{a}_{\alpha o}^2} \underset{\sim}{\underline{E}} \cdot \underset{\sim}{\mu}_o +$$

$$+ \frac{18\,\epsilon\,\underline{e}_\alpha}{(2\epsilon + \epsilon_o)^2 (\underline{kT})^2\, \underline{a}_{\alpha o}^3} \left(\underset{\sim}{\underline{E}} \cdot \underset{\sim}{\mu}_o \;\; \underset{\sim}{\mu}_o \cdot \underset{\sim}{\underline{1}}_{\underline{r}_{\alpha o}} - \frac{1}{3} \mu_o^2\, \underset{\sim}{\underline{E}} \cdot \underset{\sim}{\underline{1}}_{\underline{r}_{\alpha o}}\right) +$$

$$- \frac{6\,\underline{D}_\alpha\,\underline{e}_\alpha^2}{(\underline{D}_o + \underline{D}_\alpha)(2\epsilon + \epsilon_o)(\underline{kT})^2} \left[1 - \left(\frac{\underline{B}_o}{\underline{a}_{\alpha o}}\right)^3\right] \frac{1}{\underline{a}_{\alpha o}^2} \left(\underset{\sim}{\underline{E}} \cdot \underset{\sim}{\underline{1}}_{\underline{r}_{\alpha o}} \underset{\sim}{\mu}_o \cdot \underset{\sim}{\underline{1}}_{\underline{r}_{\alpha o}} - \frac{1}{3} \underset{\sim}{\underline{E}} \cdot \underset{\sim}{\mu}_o\right) .$$

The second boundary condition on $\underline{g}_{o\alpha}^{(2,1)}(\underset{\sim}{\underline{q}}_o ; \underset{\sim}{\underline{q}}_\alpha)$ is still given by equation 136.

In order to solve equation 144 we assume that a solution exists, expressible in the form

$$\underline{g}_{o\alpha}^{(2,1)} = \underline{f}^{(0)} + \underset{\sim}{\underline{E}} \cdot \nabla_{\underset{\sim}{\underline{r}}_{\alpha o}} \underline{f}^{(1)} + \underline{g}^{(1)}\, \underset{\sim}{\underline{E}} \cdot \underset{\sim}{\mu}_o + \qquad (146)$$

$$+ \quad \underline{f}^{(2)} \left(\underset{\sim}{E} \cdot \underset{\sim}{1}_{\underline{r}_{\alpha o}} \, \underset{\sim}{\mu}_o \cdot \underset{\sim}{1}_{\underline{r}_{\alpha o}} - \frac{1}{3} \, \underset{\sim}{E} \cdot \underset{\sim}{\mu}_o \right) +$$

$$+ \left(\underset{\sim}{E} \cdot \underset{\sim}{\mu}_o \, \underset{\sim}{\mu}_o \cdot \nabla_{\underset{\sim}{\underline{r}}_{\alpha o}} - \frac{1}{3} \, \mu_o^2 \, \underset{\sim}{E} \cdot \nabla_{\underset{\sim}{\underline{r}}_{\alpha o}} \right) \underline{g}^{(2)}$$

where $\underline{f}^{(0)}$, $\underline{f}^{(1)}$, $\underline{f}^{(2)}$, $\underline{g}^{(1)}$, and $\underline{g}^{(2)}$ are functions of $\underline{r}_{\alpha o}$ only. The term $\mu_o \, \underline{f}^{(2)} \left(\underset{\sim}{E} \cdot \underset{\sim}{1}_{\underline{r}_{\alpha o}} \, \underset{\sim}{1}_{\underline{r}_{\alpha o}} \cdot \underset{\sim}{1}_{\underline{R}_o} - \frac{1}{3} \underset{\sim}{E} \cdot \underset{\sim}{1}_{\underline{R}_o} \right)$ is a surface harmonic of degree one on the sphere $\underline{R}_o$ = constant; and it is also a surface harmonic of degree two on the sphere $\underline{r}_{\alpha o}$ = constant. The term $\mu_o^2 \, \frac{\partial \, \underline{g}^{(2)}}{\partial \, \underline{r}_{\alpha o}} \left(\underset{\sim}{E} \cdot \underset{\sim}{1}_{\underline{R}_o} \, \underset{\sim}{1}_{\underline{R}_o} \cdot \underset{\sim}{1}_{\underline{r}_{\alpha o}} - \frac{1}{3} \, \underset{\sim}{E} \cdot \underset{\sim}{1}_{\underline{r}_{\alpha o}} \right)$ is a surface harmonic of degree two on the sphere $\underline{R}_o$ = constant; and it is also a surface harmonic of degree one on the sphere $\underline{r}_{\alpha o}$ = constant. We now substitute equation 146 into equation 144, expand in surface harmonics, and equate the coefficients of the linearly independent surface harmonics to zero. In this manner we obtain five second order differential equations

$$\frac{\underline{d}^2 \; \underline{f}^{(0)}}{\underline{d} \;\; \underline{r}_{\alpha o}^{\;2}} + \frac{2}{\underline{r}_{\alpha o}} \, \frac{\underline{d} \; \underline{f}^{(0)}}{\underline{d} \; \underline{r}_{\alpha o}} = 0 \qquad (147)$$

$$\frac{d^2 f^{(1)}}{d\ r_{\alpha o}^2} + \frac{2}{r_{\alpha o}} \frac{d\ f^{(1)}}{d\ r_{\alpha o}} =$$

$$= \frac{D_\alpha\ e_\alpha}{(D_o + D_\alpha)\ kT} \sum_{n=4}^{+\infty} \frac{d_{o\alpha,n}}{kT} \left(\frac{a_{\alpha o}}{r_{\alpha o}}\right)^n \left[1 + \frac{n(n+1)}{2(n+3)} \left(\frac{a_{\alpha o}}{r_{\alpha o}}\right)^3 + - \frac{n}{2} \left(\frac{B_o}{r_{\alpha o}}\right)^3\right]$$

$$\frac{d^2 g^{(1)}}{d\ r_{\alpha o}^2} + \frac{2}{r_{\alpha o}} \frac{d\ g^{(1)}}{d\ r_{\alpha o}} - \lambda_\alpha^2\ g^{(1)} =$$

$$= - \frac{3\epsilon \lambda_\alpha^2}{(2\epsilon + \epsilon_o) kT} \left[1 + \sum_{n=4}^{+\infty} \frac{d_{o\alpha,n}}{kT} \left(\frac{a_{\alpha o}}{r_{\alpha o}}\right)^n \left(1 - \frac{n(n-1)}{\lambda_\alpha^2\ r_{\alpha o}^2}\right)\right] +$$

$$+ \frac{3\ D_\alpha\ e_\alpha^2}{(D_o + D_\alpha)(2\epsilon + \epsilon_o)(kT)^2} \left(\frac{1}{r_{\alpha o}^6} \left[a_{\alpha o}^3 - 2 B_o^3\right] + - \frac{\lambda_\alpha^2}{6\ r_{\alpha o}^4} \left[a_{\alpha o}^3 - B_o^3\right]\right)$$

$$\frac{d^2\, f^{(2)}}{d\, r_{\alpha o}^{\,2}} + \frac{2}{r_{\alpha o}}\,\frac{d\, f^{(2)}}{d\, r_{\alpha o}} - \left(\frac{6}{r_{\alpha o}^{\,2}} + \lambda_\alpha^2\right) f^{(2)} =$$

$$= \frac{3\, D_\alpha\, e_\alpha^{\,2}}{(D_o + D_\alpha)(2\epsilon + \epsilon_o)(kT)^2}\left(\frac{3}{r_{\alpha o}^{\,3}} + \frac{3}{2\, r_{\alpha o}^{\,6}}\left[a_{\alpha o}^{\,3} - 2\, B_o^{\,3}\right] + \right.$$

$$\left. - \frac{\lambda_\alpha^2}{2\, r_{\alpha o}^{\,4}}\left[a_{\alpha o}^{\,3} - B_o^{\,3}\right]\right)$$

$$\frac{d^2\, g^{(2)}}{d\, r_{\alpha o}^{\,2}} + \frac{2}{r_{\alpha o}}\,\frac{d\, g^{(2)}}{d\, r_{\alpha o}} - 3\,\lambda_\alpha^2\, g^{(2)} =$$

$$= - \frac{27\,\epsilon\, e_\alpha\,\lambda_\alpha^2}{(2\epsilon + \epsilon_o)^2\,(kT)^2}\,\frac{1}{r_{\alpha o}}\ .$$

Several inconsequential constants of integration have been neglected. Solutions of these five differential equations are

$$f^{(0)} = N^{(0)}\,\frac{1}{r_{\alpha o}} + M^{(0)} \tag{148}$$

$$f^{(1)} = N^{(1)} \frac{1}{r_{\alpha o}} + M^{(1)} +$$

$$+ \frac{D_\alpha e_\alpha}{(D_o + D_\alpha) kT} \sum_{n=4}^{+\infty} \frac{d_{o\alpha,n}}{kT} \left[\frac{a_{\alpha o}^{n}}{(n-2)(n-3)} \frac{1}{r_{\alpha o}^{n-2}} + \right.$$

$$\left. + \frac{a_{\alpha o}^{n+3}}{2(n+3)} \frac{1}{r_{\alpha o}^{n+1}} - \frac{a_{\alpha o}^{n}}{2(n+1)} \frac{B_o^3}{r_{\alpha o}^{n+1}} \right]$$

$$g^{(1)} = Q^{(1)} \frac{1}{\lambda_\alpha r_{\alpha o}} \exp\left[-\lambda_\alpha r_{\alpha o}\right] + P^{(1)} \frac{1}{\lambda_\alpha r_{\alpha o}} \exp\left[\lambda_\alpha r_{\alpha o}\right]$$

$$+ \frac{3\epsilon}{2\epsilon + \epsilon_o} \frac{1}{kT} \left[1 + \sum_{n=4}^{+\infty} \frac{d_{o\alpha,n}}{kT} \left(\frac{a_{\alpha o}}{r_{\alpha o}} \right)^{n} \right]$$

$$+ \frac{3 D_\alpha e_\alpha^2}{12 (D_o + D_\alpha)(2\epsilon + \epsilon_o)(kT)^2} \left[\frac{1}{r_{\alpha o}^4} \left[a_{\alpha o}^3 - 2 B_o^3 \right] + \right.$$

$$\left. + \frac{\lambda_\alpha^3 a_{\alpha o}^3}{4 r_{\alpha o}} F(\lambda_\alpha r_{\alpha o}) \right]$$

$$\underline{f}^{(2)} = \underline{N}^{(2)} \left(\frac{3}{\lambda_\alpha^3 \, \underline{r}_{\alpha o}^3} + \frac{3}{\lambda_\alpha^2 \, \underline{r}_{\alpha o}^2} + \frac{1}{\lambda_\alpha \, \underline{r}_{\alpha o}} \right) \exp\left[- \lambda_\alpha \, \underline{r}_{\alpha o} \right]$$

$$+ \underline{M}^{(2)} \left(\frac{3}{\lambda_\alpha^3 \, \underline{r}_{\alpha o}^3} - \frac{3}{\lambda_\alpha^2 \, \underline{r}_{\alpha o}^2} + \frac{1}{\lambda_\alpha \, \underline{r}_{\alpha o}} \right) \exp\left[\lambda_\alpha \, \underline{r}_{\alpha o} \right]$$

$$+ \frac{3 \, \underline{D}_\alpha \, \underline{e}_\alpha^2}{(\underline{D}_o + \underline{D}_\alpha)(2\epsilon + \epsilon_o) \, (\mathbf{k}T)^2} \left[- \frac{3}{\lambda_\alpha^2 \, \underline{r}_{\alpha o}^3} + \right.$$

$$+ \frac{1}{\underline{r}_{\alpha o}^4} \left(\frac{7}{16} \, \underline{a}_{\alpha o}^3 - \frac{1}{2} \, \underline{B}_o^3 \right) - \frac{3}{64} \, \frac{\lambda_\alpha^2 \, \underline{a}_{\alpha o}^3}{\underline{r}_{\alpha o}^2} \, \underline{F}'(\lambda_\alpha \, \underline{r}_{\alpha o})$$

$$\left. + \frac{1}{64} \left(3 + \lambda_\alpha^2 \, \underline{r}_{\alpha o}^2 \right) \frac{\lambda_\alpha \, \underline{a}_{\alpha o}^3}{\underline{r}_{\alpha o}^3} \, \underline{F}(\lambda_\alpha \, \underline{r}_{\alpha o}) \right]$$

$$\underline{g}^{(2)} = \underline{Q}^{(2)} \, \frac{1}{\sqrt{3} \, \lambda_\alpha \, \underline{r}_{\alpha o}} \exp\left[- \sqrt{3} \, \lambda_\alpha \, \underline{r}_{\alpha o} \right] +$$

$$+ \underline{P}^{(2)} \, \frac{1}{\sqrt{3} \, \lambda_\alpha \, \underline{r}_{\alpha o}} \exp\left[\sqrt{3} \, \lambda_\alpha \, \underline{r}_{\alpha o} \right] +$$

$$+ \frac{9 \epsilon \, \underline{e}_\alpha}{(2\epsilon + \epsilon_o)^2 \, (\underline{kT})^2} \, \frac{1}{\underline{r}_{\alpha o}} \, .$$

The function $\underline{F}(\underline{x})$ and its derivative $\underline{F}'(\underline{x})$ are defined as follows

$$\underline{F}(\underline{x}) = -\frac{2}{\underline{x}} + e^{\underline{x}}\,\underline{Ei}(\underline{x}) - e^{-\underline{x}}\,\underline{Ei}(-\underline{x}) \tag{149}$$

$$\underline{F}'(\underline{x}) = \frac{2}{\underline{x}^2} + e^{\underline{x}}\,\underline{Ei}(\underline{x}) + e^{-\underline{x}}\,\underline{Ei}(-\underline{x})$$

in terms of the exponential integrals[32]

$$\underline{Ei}(\underline{x}) = \int_{\underline{x}}^{+\infty} \frac{1}{\underline{u}}\, e^{-\underline{u}}\, d\underline{u} \tag{150}$$

$$-\underline{Ei}(-\underline{x}) = \int_{-\infty}^{\underline{x}} \frac{1}{\underline{u}}\, e^{\underline{u}}\, d\underline{u} \quad .$$

From equation 136 the constants of integration $\underline{M}^{(0)}$, $\underline{P}^{(1)}$, $\underline{M}^{(2)}$, and $\underline{P}^{(2)}$ must each be zero. The inconsequential constant $\underline{M}^{(1)}$ can be neglected. From equation 145 the constants $\underline{N}^{(0)}$ and $\underline{Q}^{(2)}$ must each be zero; and

(151)

$$\underline{N}^{(1)} = \frac{\underline{D}_\alpha e_\alpha \underline{a}_{\alpha o}^3}{2(\underline{D}_o + \underline{D}_\alpha)\, \underline{kT}} \left[1 - \left(\frac{\underline{B}_o}{\underline{a}_{\alpha o}}\right)^3 - 3 \sum_{\underline{n}=4}^{+\infty} \frac{\underline{d}_{o\alpha,\underline{n}}}{\underline{kT}} \frac{\underline{n}+1}{(\underline{n}-3)(\underline{n}+3)} \right]$$

$$+ \frac{3\epsilon\, e_\alpha \mu_o^2}{(2\epsilon + \epsilon_o)^2 (\underline{kT})^2}$$

$$\underline{Q}^{(1)} = \frac{\underline{D}_\alpha e_\alpha^2 \lambda_\alpha \exp(\lambda_\alpha \underline{a}_{\alpha o})}{(\underline{D}_o + \underline{D}_\alpha)(2\epsilon + \epsilon_o)(\underline{kT})^2 (1 + \lambda_\alpha \underline{a}_{\alpha o})} \left[1 + \right.$$

$$\left. - \frac{1}{16} \lambda_\alpha^3 \underline{a}_{\alpha o}^3 \underline{F}(\lambda_\alpha \underline{a}_{\alpha o}) + \frac{1}{16} \lambda_\alpha^4 \underline{a}_{\alpha o}^4 \underline{F}'(\lambda_\alpha \underline{a}_{\alpha o}) \right]$$

$$\underline{N}^{(2)} = \frac{3\underline{D}_\alpha e_\alpha^2 \lambda_\alpha \exp(\lambda_\alpha \underline{a}_{\alpha o})}{(\underline{D}_o + \underline{D}_\alpha)(2\epsilon + \epsilon_o)(\underline{kT})^2} \left[\frac{1 + \frac{7}{144}\lambda_\alpha^2 \underline{a}_{\alpha o}^2}{1 + \lambda_\alpha \underline{a}_{\alpha o}} \right.$$

$$- \frac{1}{64} \lambda_\alpha^3 \underline{a}_{\alpha o}^3 \left[1 + \frac{4}{9} \lambda_\alpha^2 \underline{a}_{\alpha o}^2 \right] \underline{F}(\lambda_\alpha \underline{a}_{\alpha o})$$

$$+ \frac{4}{9} \lambda_\alpha^2 \underline{a}_{\alpha o}^2 + \frac{1}{9} \lambda_\alpha^3 \underline{a}_{\alpha o}^3$$

$$\left. + \frac{1}{64} \lambda_\alpha^4 \underline{a}_{\alpha o}^4 \left[1 + \frac{1}{9} \lambda_\alpha^2 \underline{a}_{\alpha o}^2 \right] \underline{F}'(\lambda_\alpha \underline{a}_{\alpha o}) \right] .$$

Therefore, through terms quadratic in $(1/\underline{kT})$, the non-equilibrium perturbation to the $o\alpha$-pair correlation function is

$$g_{o\alpha}^{(2,1)}(\tilde{q}_o;\tilde{q}_\alpha) = \left[-N^{(1)} \frac{1}{r_{\alpha o}^2} + \frac{D_\alpha e_\alpha}{(D_o + D_\alpha)\,kT} \sum_{n=4}^{+\infty} \frac{d_{o\alpha,n}}{kT} \left(- \frac{1}{(n-3)} \frac{a_{\alpha o}^{n}}{r_{\alpha o}^{n-1}} - \frac{(n+1)}{2(n+3)} \frac{a_{\alpha o}^{n+3}}{r_{\alpha o}^{n+2}} + \frac{1}{2} B_o^3 \frac{a_{\alpha o}^{n}}{r_{\alpha o}^{n+2}} \right) \right] \tilde{E} \cdot \tilde{1}_{r_{\alpha o}} + \left[Q^{(1)} \frac{1}{\lambda_\alpha r_{\alpha o}} \exp\left[-\lambda_\alpha r_{\alpha o}\right] + \frac{3\epsilon}{(2\epsilon + \epsilon_o)\,kT} \left\{ 1 + \sum_{n=4}^{+\infty} \frac{d_{o\alpha,n}}{kT} \left(\frac{a_{\alpha o}}{r_{\alpha o}} \right)^{n} \right\} + \frac{D_\alpha e_\alpha^2}{4(D_o + D_\alpha)(2\epsilon + \epsilon_o)(kT)^2} \left\{ \frac{1}{r_{\alpha o}^4} (a_{\alpha o}^3 - 2B_o^3) + \frac{\lambda_\alpha^3 a_{\alpha o}^3}{4\, r_{\alpha o}} F(\lambda_\alpha r_{\alpha o}) \right\} \right] \tilde{E} \cdot \tilde{\mu}_o + \qquad (152)$$

$$+ \left[\underline{N}^{(2)} \left(\frac{3}{\lambda_\alpha^3 \underline{r}_{\alpha o}^3} + \frac{3}{\lambda_\alpha^2 \underline{r}_{\alpha o}^2} + \frac{1}{\lambda_\alpha \underline{r}_{\alpha o}} \right) \exp\left[- \lambda_\alpha \underline{r}_{\alpha o} \right] + \right.$$

$$+ \frac{3\, \underline{D}_\alpha\, \underline{e}_\alpha^2}{(\underline{D}_o + \underline{D}_\alpha)(2\epsilon + \epsilon_o)(\underline{kT})^2} \left\{ - \frac{3}{\lambda_\alpha^2 \underline{r}_{\alpha o}^3} + \right.$$

$$+ \frac{1}{\underline{r}_{\alpha o}^4} \left(\frac{7}{16}\, \underline{a}_{\alpha o}^3 - \frac{1}{2}\, \underline{B}_o^3 \right) +$$

$$+ \frac{1}{64} \left(3 + \lambda_\alpha^2 \underline{r}_{\alpha o}^2 \right) \frac{\lambda_\alpha \underline{a}_{\alpha o}^3}{\underline{r}_{\alpha o}^3} \underline{F}(\lambda_\alpha \underline{r}_{\alpha o}) +$$

$$\left. \left. - \frac{3}{64} \frac{\lambda_\alpha^2 \underline{a}_{\alpha o}^3}{\underline{r}_{\alpha o}^2} \underline{F}'(\lambda_\alpha \underline{r}_{\alpha o}) \right\} \right] \left(\underset{\sim}{\underline{E}} \cdot \underset{\sim}{1}_{\underline{r}_{\alpha o}}\, \underset{\sim}{\mu}_o \cdot \underset{\sim}{1}_{\underline{r}_{\alpha o}} - \frac{1}{3} \underset{\sim}{\underline{E}} \cdot \underset{\sim}{\mu}_o \right)$$

$$- \frac{9\epsilon\, \underline{e}_\alpha}{(2\epsilon + \epsilon_o)^2 (\underline{kT})^2} \frac{1}{\underline{r}_{\alpha o}^2} \left(\underset{\sim}{\underline{E}} \cdot \underset{\sim}{\mu}_o\, \underset{\sim}{\mu}_o \cdot \underset{\sim}{1}_{\underline{r}_{\alpha o}} - \frac{1}{3} \mu_o^2\, \underset{\sim}{\underline{E}} \cdot \underset{\sim}{1}_{\underline{r}_{\alpha o}} \right)$$

where $\underline{N}^{(1)}$, $\underline{N}^{(2)}$, and $\underline{Q}^{(1)}$ are the constants of integration given by equation 151.

By simply averaging over orientations of the non-electrolyte molecule and of the electrolyte ion — see equation 49 and Appendix C — we obtain the non-equilibrium perturbation to the oα-pair position correlation function

$$g_{o\alpha}^{(2,1)}(\underset{\sim}{r}_o ; \underset{\sim}{r}_\alpha) = \left[- N^{(1)} \frac{1}{r_{\alpha o}^2} + \right. \tag{153}$$

$$+ \frac{D_\alpha e_\alpha}{(D_o + D_\alpha) kT} \sum_{n=4}^{+\infty} \frac{d_{o\alpha,n}}{kT} \left(- \frac{1}{(n-3)} \frac{a_{\alpha o}^{n}}{r_{\alpha o}^{n-1}} + \right.$$

$$\left. \left. - \frac{(n+1)}{2(n+3)} \frac{a_{\alpha o}^{n+3}}{r_{\alpha o}^{n+2}} + \frac{1}{2} B_o^3 \frac{a_{\alpha o}^{n}}{r_{\alpha o}^{n+2}} \right) \right] \underset{\sim}{E} \cdot \underset{\sim}{1}r_{\alpha o} .$$

The relaxation force $\underset{\sim}{F}_o^{(1)}(\underset{\sim}{r}_o)$ on an ideal dipolar non-electrolyte molecule can be calculated from equation 119. The oα-pair correlation function $g_{o\alpha}^{(2)}(\underset{\sim}{q}_o ; \underset{\sim}{q}_\alpha)$ is given by equations 44, 129, and 152; and the force $\underset{\sim}{F}_{\alpha o}(\underset{\sim}{q}_\alpha, \underset{\sim}{q}_o)$ is given by equation D 16. Integration yields the result

$$\underset{\sim}{F}_o^{(1)}(\underset{\sim}{r}_o) = - \frac{2\pi}{3} \underset{\sim}{E} \sum_{\alpha=1}^{\sigma} \frac{D_\alpha C_\alpha e_\alpha a_{\alpha o}^3}{(D_o + D_\alpha)} \left(1 - \frac{B_o^3}{a_{\alpha o}^3} \right) \sum_{n=4}^{+\infty} \frac{d_{o\alpha,n}}{kT} + \tag{154}$$

$$+ \frac{2\pi}{3} \underset{\sim}{E} \sum_{\alpha=1}^{\sigma} \frac{D_\alpha C_\alpha e_\alpha a_{\alpha o}^3}{(D_o + D_\alpha)} \sum_{n=4}^{+\infty} \sum_{l=4}^{+\infty} \frac{d_{o\alpha,n}}{kT} \frac{d_{o\alpha,l}}{kT} \left(\frac{1}{(n+l)} \frac{B_o^3}{a_{\alpha o}^3} + \right.$$

$$- \frac{n^2(l - 3) + n(l^2 - 6l - 3) - 6l}{(n + 3)(n + 1)(n + l - 3)}\Bigg) \quad +$$

$$- \frac{4\pi\epsilon\mu_0^2}{(2\epsilon + \epsilon_0)^2\, kT} \frac{E}{\aleph} \sum_{\alpha=1}^{\sigma} C_\alpha e_\alpha \sum_{n=4}^{+\infty} \frac{d_{o\alpha,n}}{kT} \quad +$$

$$+ \frac{\pi \mu_0^2}{(2\epsilon + \epsilon_0)^2 (kT)^2} \frac{E}{\aleph} \sum_{\alpha=1}^{\sigma} \frac{D_\alpha C_\alpha e_\alpha^3}{(D_0 + D_\alpha)\, a_{\alpha o}} \left\{\frac{B_0^3}{a_{\alpha o}^3} + G(\lambda_\alpha a_{\alpha o})\right\}.$$

The function $G(x)$ defined by

$$G(x) = \frac{13 - 3x - \frac{13}{6}x^2 - \frac{7}{6}x^3 + \frac{1}{2}x^4 e^x \, Ei(x)}{6 + 6x + \frac{8}{3}x^2 + \frac{2}{3}x^3} \qquad (155)$$

is tabulated in Appendix E. The first three terms on the right in equation 154 arise from the short range forces exerted on the non-electrolyte molecule by the surrounding electrolyte ions; and the final term results from the electrostatic interaction of the dipole with the electrolyte ions.

The kinetic force $\underset{\sim}{K}_o^{(1)}(\underset{\sim}{r}_o)$ on an ideal dipolar non-electrolyte molecule can be calculated from equation 120. When $\mathcal{R}_o \geq a_{\alpha o}$, the function $g_{o\alpha}^{(2)}(\underset{\sim}{r}_o; \underset{\sim}{r}_o + \mathcal{R}_o \underset{\sim}{1}_{r_{\alpha o}})$ is given by equations 130 and 153 and by an equation similar to equation 44 ; it vanishes when $\mathcal{R}_o < a_{\alpha o}$. In order to achieve a simple self-consistant description of the system, we shall refine our model[33] by identifying the hydrodynamic radius of the non-electrolyte molecule $\mathcal{R}_o$ with the distance of closest approach $a_{\alpha o}$. As mentioned before, the kinetic entity moving in the solvent is not necessarily the bare non-electrolyte molecule of radius b_o ; usually it is the non-electrolyte molecule plus a number of adjacent solvent molecules. This entity is approximated by a sphere of radius $\mathcal{R}_o$. This sphere is deformable; and as an electrolyte ion of species α approaches it, the solvent is squeezed out of the region between them. Thus it appears reasonable to assume the kinetic entity to be a sphere of radius $a_{\alpha o}$. Upon replacing $\mathcal{R}_o$ by $a_{\alpha o}$ and carrying out the integration we obtain the result

$$\underset{\sim}{K}_o^{(1)}(\underset{\sim}{r}_o) = \frac{2\pi}{3} \underset{\sim}{E} \sum_{\alpha=1}^{\sigma} \frac{D_\alpha C_\alpha e_\alpha a_{\alpha o}^3}{(D_o + D_\alpha)} \left(1 - \frac{B_o^3}{a_{\alpha o}^3}\right) + \tag{156}$$

$$+ \frac{2\pi}{3} \underset{\sim}{E} \sum_{\alpha=1}^{\sigma} \frac{D_\alpha C_\alpha e_\alpha a_{\alpha o}^3}{(D_o + D_\alpha)} \sum_{n=4}^{+\infty} \frac{d_{o\alpha,n}}{kT} \left(\frac{n}{n+3} - \frac{B_o^3}{a_{\alpha o}^3} \right) .$$

The first term on the right of this equation depends "only" on the geometrical properties of the cavity of "low" dielectric constant and of the electrolyte ions, whereas the last term depends on the short range intermolecular forces as well.

The electrophoretic velocity of an ideal dipolar non-electrolyte molecule can be calculated from equation 121. The function $g_{o\alpha}^{(2,0)}(r_{\alpha o})$ is given by equation 130. The result of this integration is

(157)

$$\underset{\sim}{v}_{10}(\underset{\sim}{r}_o \mid \underset{\sim}{r}_o) = \frac{2}{3\eta} \underset{\sim}{E} \sum_{\alpha=1}^{\sigma} C_\alpha e_\alpha a_{\alpha o}^2 \sum_{n=4}^{+\infty} \frac{1}{(n-2)} \frac{d_{o\alpha,n}}{kT} +$$

$$+ \frac{1}{3\eta} \underset{\sim}{E} \sum_{\alpha=1}^{\sigma} C_\alpha e_\alpha a_{\alpha o}^2 \sum_{n,l=4}^{+\infty} \frac{1}{(n+l-2)} \frac{d_{o\alpha,n}}{kT} \frac{d_{o\alpha,l}}{kT} +$$

$$+ \frac{\mu_o^2}{2\eta(2\epsilon + \epsilon_o)^2 (kT)^2} \underset{\sim}{E} \sum_{\alpha=1}^{\sigma} C_\alpha e_\alpha^3 \frac{1}{a_{\alpha o}^2} .$$

Again we have replaced $\mathcal{R}_o$ by $\underline{a}_{\alpha o}$. The first two terms on the right of this equation arise from the short range forces between the non-electrolyte molecule and the electrolyte ions of its atmosphere; and the final term results from the electrostatic interaction of the dipole with the electrolyte ions. An alternate procedure would be to solve the Navier-Stokes equation of motion, see equations 96 and 99, assuming the ideal dipolar non-electrolyte molecule to be held in a fixed orientation relative to the applied electrostatic field. After obtaining $\underset{\sim}{\underline{v}}_{|o}(\underset{\sim}{\underline{r}}_o \mid \underset{\sim}{\underline{q}}_o)$ in this manner, averaging over orientations of the non-electrolyte molecule yields equation 157.

The mobility of the ideal dipolar non-electrolyte, as obtained from equations 108, 118, 154, 156, and 157, is given by

$$\underline{U}_o = \underline{U}_o^{(I)} + \underline{U}_o^{(II)} + \underline{U}_o^{(III)} + \underline{U}_o^{(IV)} \qquad (158)$$

where

$$\underline{U}_o^{(I)} = \frac{2\pi}{3} \frac{\underline{D}_o}{\underline{kT}} \sum_{\alpha=1}^{\sigma} \frac{\underline{D}_\alpha \underline{C}_\alpha \underline{e}_\alpha}{(\underline{D}_o + \underline{D}_\alpha)} \left(\underline{a}_{\alpha o}^{3} - \underline{B}_o^{3} \right)$$

$$\underline{U}_o^{(II)} = \frac{\mu_o^2}{2\eta(2\epsilon + \epsilon_o)^2(\underline{kT})^2} \sum_{\alpha=1}^{\sigma} \underline{C}_\alpha \underline{e}_\alpha^3 \frac{1}{\underline{a}_{\alpha o}^2} +$$

$$+ \frac{\pi \mu_o^2 \underline{D}_o}{(2\epsilon + \epsilon_o)^2 (\underline{kT})^3} \sum_{\alpha=1}^{\sigma} \frac{\underline{D}_\alpha \underline{C}_\alpha \underline{e}_\alpha^3}{(\underline{D}_o + \underline{D}_\alpha)} \frac{1}{\underline{a}_{\alpha o}} \left(\frac{\underline{B}_o^3}{\underline{a}_{\alpha o}^3} + \underline{G}(\lambda_\alpha \underline{a}_{\alpha o}) \right)$$

$$\underline{U}_o^{(III)} = \frac{2}{3\eta} \sum_{\alpha=1}^{\sigma} \underline{C}_\alpha \underline{e}_\alpha \underline{a}_{\alpha o}^2 \sum_{\underline{n}=4}^{+\infty} \frac{1}{(\underline{n} - 2)} \frac{\underline{d}_{o\alpha,\underline{n}}}{\underline{kT}} +$$

$$- 2\pi \frac{\underline{D}_o}{\underline{kT}} \sum_{\alpha=1}^{\sigma} \frac{\underline{D}_\alpha \underline{C}_\alpha \underline{e}_\alpha}{(\underline{D}_o + \underline{D}_\alpha)} \underline{a}_{\alpha o}^3 \sum_{\underline{n}=4}^{+\infty} \frac{1}{(\underline{n} + 3)} \frac{\underline{d}_{o\alpha,\underline{n}}}{\underline{kT}} +$$

$$+ \frac{1}{3\eta} \sum_{\alpha=1}^{\sigma} \underline{C}_\alpha \underline{e}_\alpha \underline{a}_{\alpha o}^2 \sum_{\underline{n},\underline{l}=4}^{+\infty} \frac{1}{(\underline{n} + \underline{l} - 2)} \frac{\underline{d}_{o\alpha,\underline{n}}}{\underline{kT}} \frac{\underline{d}_{o\alpha,\underline{l}}}{\underline{kT}} +$$

$$+ \frac{2\pi}{3} \frac{\underline{D}_o}{\underline{kT}} \sum_{\alpha=1}^{\sigma} \frac{\underline{D}_\alpha \underline{C}_\alpha \underline{e}_\alpha}{(\underline{D}_o + \underline{D}_\alpha)} \underline{a}_{\alpha o}^3 \sum_{\underline{n},\underline{l}=4}^{+\infty} \frac{\underline{d}_{o\alpha,\underline{n}}}{\underline{kT}} \frac{\underline{d}_{o\alpha,\underline{l}}}{\underline{kT}} \left(\frac{1}{(\underline{n} + \underline{l})} \frac{\underline{B}_o^3}{\underline{a}_{\alpha o}^3} + \right.$$

$$\left. - \frac{\underline{n}^2(\underline{l} - 3) + \underline{n}(\underline{l}^2 - 6\underline{l} - 3) - 6\underline{l}}{(\underline{n} + 3)(\underline{n} + \underline{l})(\underline{n} + \underline{l} - 3)} \right)$$

$$\underline{U}_o^{(IV)} = - \frac{4\pi\epsilon\mu_o^2 \underline{D}_o}{(2\epsilon + \epsilon_o)^2(\underline{kT})^2} \sum_{\alpha=1}^{\sigma} \underline{c}_\alpha \underline{e}_\alpha \sum_{\underline{n}=4}^{+\infty} \frac{\underline{d}_{o\alpha,\underline{n}}}{\underline{kT}}$$

In this limiting "law" only terms linear in the electrolyte concentration appear. Our result is independent of the concentration of the non-electrolyte. The first term $\underline{U}_o^{(I)}$ is a kinetic or pressure term which results from collisions of the electrolyte ions with the non-electrolyte molecule considered to be an uncharged rigid sphere of low dielectric constant. The second term $\underline{U}_o^{(II)}$ results from the long range electrostatic ion-dipole interactions between the non-electrolyte molecule and the electrolyte ions in its atmosphere. It consists of two parts: the first arises through the electrophoretic effect; the second arises from the relaxation force. The third term $\underline{U}_o^{(III)}$ results from the short range forces between the non-electrolyte molecule and the electrolyte ions. It consists of four parts: the first and third arise through the electrophoretic effect; the fourth arises from the relaxation force; and the second arises from both the relaxation force and the kinetic force. Finally, the fourth term $\underline{U}_o^{(IV)}$ is a cross-term resulting from both short range forces and long range ion-dipole forces. It arises from the relaxation force.

The mass transport ratio of an ideal dipolar non-electrolyte, as obtained from equations 109, 117, 118, 154, 156, and 157, is given by

$$t_o^{(m)} = t_o^{(m,\, I)} + t_o^{(m,\, II)} + t_o^{(m,\, III)} + t_o^{(m,\, IV)} \tag{159}$$

where

$$t_o^{(m,\, I)} = \frac{2\pi m_o}{3} D_o C_o \frac{\sum_{\alpha=1}^{\sigma} \frac{D_\alpha C_\alpha e_\alpha}{(D_o + D_\alpha)} \left[a_{\alpha o}^3 - B_o^3\right]}{\sum_{\alpha=1}^{\sigma} D_\alpha C_\alpha e_\alpha^2}$$

$$t_o^{(m,\, II)} = \frac{m_o\, \mu_o^2}{2\eta(2\epsilon + \epsilon_o)^2(kT)} C_o \frac{\sum_{\alpha=1}^{\sigma} C_\alpha e_\alpha^3 \frac{1}{a_{\alpha o}^2}}{\sum_{\alpha=1}^{\sigma} D_\alpha C_\alpha e_\alpha^2} +$$

$$+ \frac{m_o \pi \mu_o^2}{(2\epsilon + \epsilon_o)^2(kT)^2} D_o C_o \frac{\sum_{\alpha=1}^{\sigma} \frac{D_\alpha C_\alpha e_\alpha^3}{(D_o + D_\alpha)} \frac{1}{a_{\alpha o}} \left(\frac{B_o^3}{a_{\alpha o}^3} + G(\lambda_\alpha a_{\alpha o})\right)}{\sum_{\alpha=1}^{\sigma} D_\alpha C_\alpha e_\alpha^2}$$

$$\underline{t}_o(\underline{m}, III) = \frac{2\underline{m}_o \underline{kT}}{3\eta} \underline{C}_o \frac{\sum_{\alpha=1}^{\sigma} \underline{C}_\alpha \underline{e}_\alpha \underline{a}_{\alpha o}^2 \sum_{\underline{n}=4}^{+\infty} \frac{1}{(\underline{n}-2)} \frac{\underline{d}_{o\alpha,\underline{n}}}{\underline{kT}}}{\sum_{\alpha=1}^{\sigma} \underline{D}_\alpha \underline{C}_\alpha \underline{e}_\alpha^2} +$$

$$- 2\pi \underline{m}_o \underline{D}_o \underline{C}_o \frac{\sum_{\alpha=1}^{\sigma} \frac{\underline{D}_\alpha \underline{C}_\alpha \underline{e}_\alpha}{(\underline{D}_o + \underline{D}_\alpha)} \underline{a}_{\alpha o}^3 \sum_{\underline{n}=4}^{+\infty} \frac{1}{(\underline{n}+3)} \frac{\underline{d}_{o\alpha,\underline{n}}}{\underline{kT}}}{\sum_{\alpha=1}^{\sigma} \underline{D}_\alpha \underline{C}_\alpha \underline{e}_\alpha^2} +$$

$$+ \frac{\underline{m}_o \underline{kT}}{3\eta} \underline{C}_o \frac{\sum_{\alpha=1}^{\sigma} \underline{C}_\alpha \underline{e}_\alpha \underline{a}_{\alpha o}^2 \sum_{\underline{n},\underline{l}=4}^{+\infty} \frac{1}{(\underline{n}+\underline{l}-2)} \frac{\underline{d}_{o\alpha,\underline{n}}}{\underline{kT}} \frac{\underline{d}_{o\alpha,\underline{l}}}{\underline{kT}}}{\sum_{\alpha=1}^{\sigma} \underline{D}_\alpha \underline{C}_\alpha \underline{e}_\alpha^2} +$$

$$+ \frac{2\pi \underline{m}_o}{3} \underline{D}_o \underline{C}_o \frac{\sum_{\alpha=1}^{\sigma} \frac{\underline{D}_\alpha \underline{C}_\alpha \underline{e}_\alpha \underline{a}_{\alpha o}^3}{(\underline{D}_o + \underline{D}_\alpha)} \sum_{\underline{n},\underline{l}=4}^{+\infty} \frac{\underline{d}_{o\alpha,\underline{n}}}{\underline{kT}} \frac{\underline{d}_{o\alpha,\underline{l}}}{\underline{kT}} \left(\frac{1}{(\underline{n}+\underline{l})} \frac{\underline{B}_o^3}{\underline{a}_{\alpha o}^3} - \frac{\underline{n}^2(\underline{l}-3) + \underline{n}(\underline{l}^2 - 6\underline{l} - 3) - 6\underline{l}}{(\underline{n}+3)(\underline{n}+\underline{l})(\underline{n}+\underline{l}-3)} \right)}{\sum_{\alpha=1}^{\sigma} \underline{D}_\alpha \underline{C}_\alpha \underline{e}_\alpha^2}$$

$$\underline{t}_o^{(\underline{m},\,IV)} = -\frac{4\pi \underline{m}_o \epsilon \mu_o^2}{(2\epsilon + \epsilon_o)^2 \underline{kT}} \underline{D}_o \underline{C}_o \frac{\sum_{\alpha=1}^{\sigma} \underline{C}_\alpha \underline{e}_\alpha \sum_{\underline{n}=4}^{+\infty} \frac{\underline{d}_{o\alpha,\underline{n}}}{\underline{kT}}}{\sum_{\alpha=1}^{\sigma} \underline{D}_\alpha \underline{C}_\alpha \underline{e}_\alpha^2} .$$

The various contributions to $\underline{t}_o^{(\underline{m})}$ have been arranged in the same way that we arranged the contributions to $\underline{U}_o$. In this limiting "law" terms linear in the non-electrolyte concentration appear. Our result is, however, independent of the electrolyte concentration.

Before comparing our limiting "laws" with experimental results, we shall consider some approximations valid for large non-electrolyte molecules. From Stokes' Equation, equation 73, we have $\underline{D}_o = \underline{kT}/6\pi\eta\mathcal{R}_o$ and $\underline{D}_\alpha = \underline{kT}/6\pi\eta\mathcal{R}_\alpha$ where $\mathcal{R}_o$ and $\mathcal{R}_\alpha$ are the hydrodynamic radii of the non-electrolyte molecules of species o and of the electrolyte ions of species α respectively. Therefore as a useful approximation we can neglect $\underline{D}_o$ relative to $\underline{D}_\alpha$ when the non-electrolyte molecules are considerably larger than the electrolyte ions, i.e., when $\mathcal{R}_o \gg \mathcal{R}_\alpha$ or when $\underline{b}_o \gg \underline{b}_\alpha$. As an example: protein molecules have translational diffusion constants of the order 10^{-7} cm^2/sec

whereas simple electrolyte ions have translational diffusion constants of the order 10^{-5} cm^2/sec.

Using equations 73, 81, and 132, we find that

$$\lambda_\alpha^2 = \frac{2\,\underline{D}_o^{(rot)}}{\underline{D}_o + \underline{D}_\alpha} = \frac{3\,\mathcal{R}_\alpha}{2\,\mathcal{R}_o^2\,(\mathcal{R}_o + \mathcal{R}_\alpha)} .$$

Consequently, for extremely large non-electrolyte molecules $\lambda_\alpha \underline{a}_{\alpha o}$ can be neglected relative to unity; and the function $\underline{G}(\lambda_\alpha \underline{a}_{\alpha o})$, appearing in equations 158 and 159, becomes simply 13/6. The same result is obtained if we assign to $\underline{D}_o^{(rot)}$ the value zero. This implies that $\zeta_o^{(rot)}$, the "inner frictional constant" of the non-electrolyte molecule, is infinite; and that the non-electrolyte molecule is being "held" in a fixed orientation relative to the applied electrostatic field.

Quite naturally the equations given here are simplified considerably if we calculate the oα-pair correlation function $\underline{g}_{o\alpha}^{(2)}(\underline{q}_o;\underline{q}_\alpha)$ with the non-electrolyte molecule in a fixed orientation relative to the applied electrostatic field. With this restriction $\underline{g}_{o\alpha}^{(2,1)}(\underline{q}_o;\underline{q}_\alpha)$ is a solution of equation 93, the equation of continuity in oα-pair position space. For an ideal dipolar non-electrolyte we obtain — in place of equation 152 — the result

$$g_{o\alpha}^{(2,1)}(q_o; q_\alpha) = \left[-N^{(1)} \frac{1}{r_{\alpha o}^2} + \right. \tag{160}$$

$$+ \frac{D_\alpha e_\alpha}{(D_o + D_\alpha) kT} \sum_{n=4}^{+\infty} \frac{d_{o\alpha,n}}{kT} \left(- \frac{1}{(n-3)} \frac{a_{\alpha o}^{n}}{r_{\alpha o}^{n-1}} + \right.$$

$$\left. \left. - \frac{(n+1)}{2(n+3)} \frac{a_{\alpha o}^{n+3}}{r_{\alpha o}^{n+2}} + \frac{1}{2} B_o^3 \frac{a_{\alpha o}^{n}}{r_{\alpha o}^{n+2}} \right) \right] E \cdot 1 r_{\alpha o} +$$

$$+ \left[\frac{D_\alpha e_\alpha^2}{(D_o + D_\alpha)(2\epsilon + \epsilon_o)(kT)^2} \left(\frac{1}{r_{\alpha o}} + \frac{1}{4 r_{\alpha o}^4} \left[a_{\alpha o}^3 - 2 B_o^3 \right] \right) \right] E \cdot \mu_o +$$

$$+ \left[\frac{3 D_\alpha e_\alpha^2}{2(D_o + D_\alpha)(2\epsilon + \epsilon_o)(kT)^2} \left(- \frac{1}{r_{\alpha o}} + \frac{a_{\alpha o}^2}{r_{\alpha o}^3} + \right. \right.$$

$$\left. \left. + \frac{1}{2 r_{\alpha o}^4} \left[a_{\alpha o}^3 - 2 B_o^3 \right] \right) \right] \left(E \cdot 1 r_{\alpha o} \, \mu_o \cdot 1 r_{\alpha o} - \frac{1}{3} E \cdot \mu_o \right)$$

where

$$N^{(1)} = \frac{D_\alpha e_\alpha a_{\alpha o}^3}{2(D_o + D_\alpha) kT} \left[1 - \left(\frac{B_o}{a_{\alpha o}} \right)^3 - 3 \sum_{n=4}^{+\infty} \frac{d_{o\alpha,n}}{kT} \frac{(n+1)}{(n-3)(n+3)} \right].$$

Using this result, the value obtained for the kinetic force $\underset{\sim}{\underline{K}}_o^{(1)}(\underset{\sim}{\underline{r}}_o)$ is the same as that given previously, equation 156. However, for the relaxation force $\underset{\sim}{\underline{F}}_o^{(1)}(\underset{\sim}{\underline{r}}_o)$, a result slightly different from that given by equation 154 is obtained: the third term on the right hand side of equation 154 has been lost; and, of course, $\underline{G}(\lambda_\alpha \underline{a}_{\alpha o})$ appearing in the fourth term has become 13/6. With these changes, the mobility $\underline{U}_o$ is given by equation 158 with $\underline{G}(\lambda_\alpha \underline{a}_{\alpha o})$ equal to 13/6 in $\underline{U}_o^{(II)}$ and with the cross-term $\underline{U}_o^{(IV)}$ omitted; and the mass transport ratio $\underline{t}_o^{(\underline{m})}$ is given by equation 159 with $\underline{G}(\lambda_\alpha \underline{a}_{\alpha o})$ equal to 13/6 in $\underline{t}_o^{(\underline{m}, II)}$ and with the cross-term $\underline{t}_o^{(\underline{m}, IV)}$ omitted. Expressions similar to these just mentioned have been given previously by Kirkwood and Squires.[34] However, they expanded the $o\alpha$-pair position space equation of continuity in terms of charging parameters of the ions and then they neglected certain higher order terms before solving. They set $\underline{B}_o$ equal to zero; and, in addition, they neglected effects due to the kinetic force.

V. Electrolytic Transport of Non-electrolytes: Comparison of Experimental and Theoretical Mass Transport Ratios

Experimental Mass Transport Ratios

The transport of non-electrolytes contained in electrolyte solutions undergoing electrolysis has received little attention. In fact, early experimenters assumed the added non-electrolyte to be uninfluenced by the passage of the electric current.

The older experimental procedures were essentially that of a Hittorf measurement, the non-electrolyte being added to the electrolyte solution. In 1900 Nernst[35] suggested that the relative degrees of solvation of ions could be determined by measuring the amount of solvent transferred from one electrode to the other during electrolysis; and that this transfer of solvent would be apparent if the solution contained a small amount of an electrically inert non-electrolyte, which would not migrate with the current and which would be increased in concentration at one electrode and decreased at the other, as a result of the solvent transference. Measurements made by Nernst and his associates were, however, inconclusive.

Successful measurements of this type were carried out

by Washburn and Millard.[36,37] They found that "the electrolysis of a solution of potassium, sodium, lithium, or cesium chloride, containing a non-electrolyte (raffinose) at low concentration, is attended by an increase in the concentration of the non-electrolyte at the anode and a corresponding decrease at the cathode, but no change in concentration takes place in the absence of the electrolyte."

Taylor and her associates[38,39] performed measurements, on sodium chloride solutions, similar to those carried out by Washburn; using, however, urea as the reference substance. An increase in the concentration of the non-electrolyte at the anode and a corresponding decrease at the cathode was again observed.

Of two possible interpretations: (1) the non-electrolyte is carried by the electrolyte from cathode to anode, or (2) water is carried by the electrolyte from anode to cathode during the passage of the current, these investigators chose the second. They calculated the number of moles of water transported per Faraday from anode to cathode, $\underline{n}_{\underline{w}}$.

In 1948 Hale and De Vries[40], who studied a series of tetraalkylammonium iodides with maltose, acetone, and ethyl acetate as added non-electrolytes; found that for a given electrolyte and concentration, $\underline{n}_{\underline{w}}$ varies with the non-electrolyte added. This result shows that the non-electrolyte itself must interact with and be transported by the ions.

Table I summarizes the aforementioned non-electrolyte transport data obtained by measurements of the Hittorf type. Values are given for the water transport $\underline{n}_w$ calculated by assuming the added non-electrolyte to be electrically inert; and for $\underline{t}_0^{(\underline{m})}/\underline{C}_0\,\underline{m}_0$, the mass transport ratio of the non-electrolyte $\underline{t}_0^{(\underline{m})}$ divided by its mass density $\underline{C}_0\,\underline{m}_0$, calculated by assuming the water to be electrically inert. In order to avoid large exponential numbers, we have expressed $\underline{t}_0^{(\underline{m})}/\underline{C}_0\,\underline{m}_0$ in units of $(cm^3/\text{Faraday})$.

More recent measurements have been of the more accurate moving boundary type. In the <u>moving boundary method</u>[41] for the measurement of non-electrolyte transport, a sharp boundary is formed within an electrophoresis cell between a solution of an electrolyte and the same solution to which a non-electrolyte has been added. The movement of this boundary, on the passage of an electric current, is then observed by the schlieren method. Diffusion at the boundary proceeds essentially independently of the current; and the boundary soon loses the sharpness with which it was initially formed. Nevertheless, its displacement and hence its velocity, with respect to the solvent, can be determined with precision. This velocity is identified with $\underline{u}_0 - \underline{u}_\tau = \underline{J}_0^{(\mathrm{rel}\,\tau)}/\underline{C}_0$, the mean local velocity of the non-electrolyte molecules relative to the mean local velocity of the solvent. The

Table I

Non-electrolyte Transport by the

Hittorf Method, at 25° C.

Non-electrolyte	Wt.%	Electrolyte	Molarity	n_w	$t_o^{(m)}/C_o m_o$ (cm^3 / Faraday)
Raffinose	4.7%	LiCl	1.22	1.5	- 28.8
Raffinose	4.4%	NaCl	1.15	0.82	- 16.2
Raffinose	4.4%	KCl	1.16	0.60	- 11.7
Raffinose	2.9%	CsCl	1.00	0.36	- 7.0
Urea[a/]	0.60%	NaCl	0.5	1.415	- 26.4
Urea[a/]	0.55%	NaCl	1.0	1.063	- 20.5
Urea	0.60%	NaCl	0.5	1.081	- 20.2
Maltose	3.4%	$(CH_3)_4NI$	0.1	4.2	- 76
Maltose	3.3%	$(C_2H_5)_4NI$	0.1	4.9	- 93
Maltose	3.3%	$(C_3H_7)_4NI$	0.1	3.6	- 65
Acetone	0.1%	$(CH_3)_4NI$	0.1	6.1	-110
Ethyl acetate	1.8%	$(CH_3)_4NI$	0.1	20.0	-360
Ethyl acetate	2.0%	$(C_2H_5)_4NI$	0.1	20.0	-360
Ethyl acetate	1.0%	$(C_4H_9)_4NI$	0.06	35.0	-631

a/These two experiments were performed at 0°C.

electric current density $\underset{\sim}{I}$ within the electrophoresis cell can be calculated from the current flowing through the cell and the cross-sectional area of the cell. Thus

$$t_0^{(m)} / C_0 m_0 = \underset{\sim}{J}_0^{(rel\,\tau)} / C_0 \underset{\sim}{I} = (\underset{\sim}{u}_0 - \underset{\sim}{u}_\tau) / \underset{\sim}{I} , \qquad (161)$$

the mass transport ratio of the non-electrolyte $t_0^{(m)}$ divided by its mass density $C_0\ m_0$, can readily be obtained. If A is the cross-sectional area of the electrophoresis cell and if t is the time during which the current passes, then $t_0^{(m)} / C_0\ m_0$ may conveniently be written in terms of the volume swept through by the boundary $A\ (\underset{\sim}{u}_0 - \underset{\sim}{u}_\tau)\ t$ divided by the electric charge passed through the cell $\underset{\sim}{I}\ A\ t$. As before, we shall express $t_0^{(m)} / C_0\ m_0$ in units of $(cm^3/\text{Faraday})$.

The moving boundary method possesses notable advantages over the older Hittorf procedure. Experiments of much greater accuracy can be performed with but a fraction of the time and effort.

Extensive moving boundary measurements have been carried out by Longsworth[42] who studied a series of chlorides with urea, resorcinol, mannitol, and raffinose as the added non-electrolytes. A summary of his experimental results is given in Table II. Values are given for both n_w and

Table II

Non-electrolyte Transport by the

Moving Boundary Method, at 0.5° C.

Non-electrolyte	Wt. %	Electrolyte	Molarity	n_w	$t_o^{(m)} / C_o\ m_o$ (cm^3/Faraday)
Urea	2.0%	LiCl	0.2	1.10	- 19.8
Urea	4.0%	LiCl	0.2	1.04	- 18.8
Resorcinol	2.0%	LiCl	0.2	1.88	- 33.8
Mannitol	2.0%	LiCl	0.2	1.19	- 21.4
Raffinose	1.5%	LiCl	0.1	1.97	- 35.5
Raffinose	1.0%	LiCl	0.2	1.86	- 33.5
Raffinose	1.0%	LiCl	0.2	1.83	- 33.0
Raffinose	2.0%	LiCl	0.2	1.72	- 31.0
Raffinose	2.0%	LiCl	1.0	1.18	- 21.2
Raffinose	1.5%	NaCl	0.2	1.06	- 19.05
Raffinose	1.0%	NaCl	0.5	0.93	- 16.8
Raffinose	1.5%	NaCl	1.0	0.80	- 14.3
Raffinose	1.5%	NaCl	1.0	0.79	- 14.1
Raffinose	1.0%	KCl	0.2	0.59	- 10.65
Raffinose	1.5%	KCl	1.0	0.47	- 8.5
Raffinose	1.5%	KCl	1.0	0.42	- 7.6
Raffinose	1.0%	NH_4Cl	0.2	0.77	- 13.8
Raffinose	1.5%	NH_4Cl	1.0	0.46	- 8.2

$\underline{t}_o^{(\underline{m})} / \underline{C}_o \underline{m}_o$, the precision of the measurements being about 1 cm^3/ Faraday. This important data proves that the added non-electrolyte is not motionless during the passage of the current. Moreover, for a given electrolyte and for given concentrations, different non-electrolytes move with different velocities. This fact can easily be seen from Figure 1 where values of $-\underline{t}_o^{(\underline{m})} / \underline{C}_o \underline{m}_o$ from Table II in which 0.2 N LiCl is the electrolyte are plotted as ordinate against the weight percent of the non-electrolyte as abscissa. The effect of increasing the concentration of the added non-electrolyte appears to be relatively small, at these low concentrations.

Other moving boundary work on non-electrolyte transport has been described by Janssen.[43] Since his experiments have been criticized by Longsworth, we shall just summarize them qualitatively. Janssen studied solutions of the sodium halides with sucrose as the added non-electrolyte. He found that the uncharged molecules of sucrose move toward the anode, in the electric field; and that with the halides of sodium the velocity of the sucrose increases in the order NaF, NaCl, and NaI.

Finally, Kirkwood and Squires,[44] using a modified moving boundary method in which the quantity of non-electrolyte crossing a datum plane in the apparatus was determined analytically, studied the transport of glycine during the electrolysis of barium chloride solutions. Their data is summarized in Table III. For this method the uncertainty

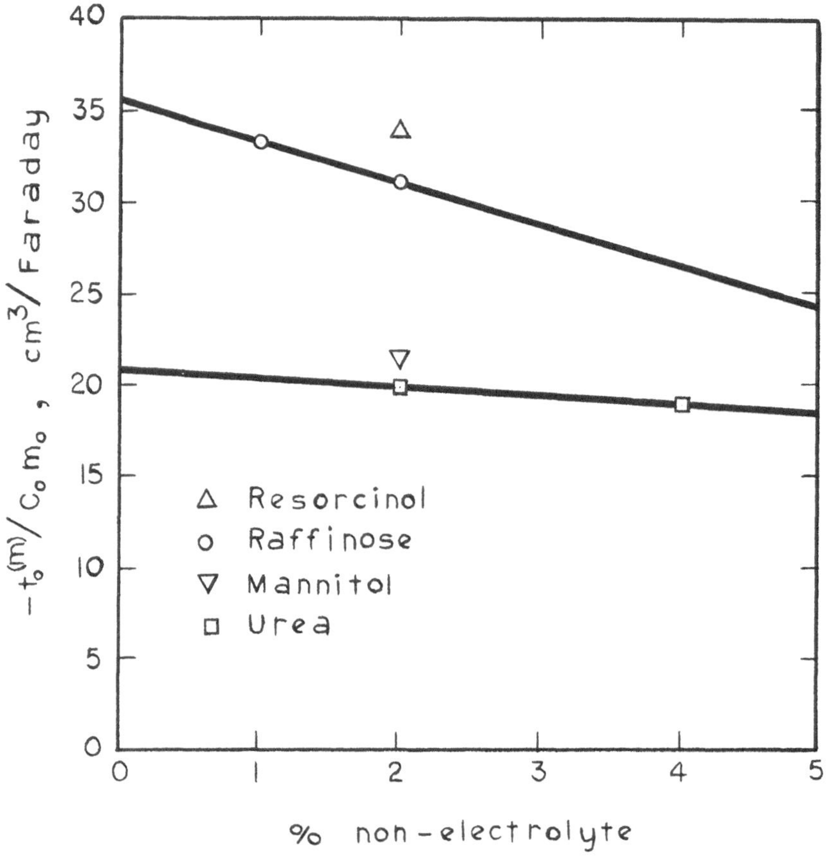

Figure 1. The Variation of $-\underline{t}_o^{(\underline{m})}/\underline{C}_o\,\underline{m}_o$ in 0.2 N LiCl Solutions with the Nature and Concentration of the Added Non-electrolyte.

of a single measurement ranges from 10% to 20%. Longsworth's method, relying on an optical determination of the position of the moving boundary, is far superior.

Table III

Transport of Glycine During the Electrolysis of $BaCl_2$ Solutions, at 25°C.

Molarity of Glycine	Molarity of $BaCl_2$	$\underline{t}_0^{(\underline{m})}/\underline{C}_0\,\underline{m}_0$ (cm^3/Faraday)
0.98	0.50	+ 51
0.93	0.25	+ 58
0.98	0.10	+ 70

Theoretical Mass Transport Ratios

In Section IV of this paper we obtained the "infinite dilution limiting law" for the mass transport ratio of an ideal dipolar non-electrolyte. This was given as equation 159. For a strong electrolyte of two ionic species, called species α and species β, we have then

$$\underline{t}_0^{(\underline{m})}/\underline{C}_0\,\underline{m}_0 = \underline{t}_0^{(\underline{m},\,I)}/\underline{C}_0\,\underline{m}_0 + \underline{t}_0^{(\underline{m},\,II)}/\underline{C}_0\,\underline{m}_0 + \tag{162}$$

$$+ \quad \underline{t}_o^{(\underline{m},\, III)} / \underline{C}_o\, \underline{m}_o \quad + \quad \underline{t}_o^{(\underline{m},\, IV)} / \underline{C}_o\, \underline{m}_o$$

with

$$\underline{t}_o^{(\underline{m},\, I)} / \underline{C}_o\, \underline{m}_o = \frac{2\pi}{3\,\underline{e}} \; \frac{\underline{D}_o}{(\underline{D}_\alpha \underline{Z}_\alpha - \underline{D}_\beta \underline{Z}_\beta)} \left[\frac{\underline{D}_\alpha}{\underline{D}_o + \underline{D}_\alpha} \left(\underline{a}_{\alpha o}^{\,3} - \underline{B}_o^{\,3} \right) + \right.$$

$$\left. - \frac{\underline{D}_\beta}{\underline{D}_o + \underline{D}_\beta} \left(\underline{a}_{\beta o}^{\,3} - \underline{B}_o^{\,3} \right) \right]$$

$$\underline{t}_o^{(\underline{m},\, II)} / \underline{C}_o\, \underline{m}_o = \frac{\mu_o^{\,2}\, \underline{e}}{2\eta\,(2\epsilon + \epsilon_o)^2\, \underline{kT}\, (\underline{D}_\alpha \underline{Z}_\alpha - \underline{D}_\beta \underline{Z}_\beta)} \left[\frac{\underline{Z}_\alpha^{\,2}}{\underline{a}_{\alpha o}^{\,2}} - \frac{\underline{Z}_\beta^{\,2}}{\underline{a}_{\beta o}^{\,2}} \right] +$$

$$+ \frac{\pi\, \mu_o^{\,2}\, \underline{e}}{(2\epsilon + \epsilon_o)^2\, (\underline{kT})^2} \; \frac{\underline{D}_o}{(\underline{D}_\alpha \underline{Z}_\alpha - \underline{D}_\beta \underline{Z}_\beta)} \left[\frac{\underline{D}_\alpha}{\underline{D}_o + \underline{D}_\alpha} \, \frac{\underline{Z}_\alpha^{\,2}}{\underline{a}_{\alpha o}} \left(\frac{\underline{B}_o^{\,3}}{\underline{a}_{\alpha o}^{\,3}} + \underline{G}(\lambda_\alpha\, \underline{a}_{\alpha o}) \right) \right.$$

$$\left. - \frac{\underline{D}_\beta}{\underline{D}_o + \underline{D}_\beta} \, \frac{\underline{Z}_\beta^{\,2}}{\underline{a}_{\beta o}} \left(\frac{\underline{B}_o^{\,3}}{\underline{a}_{\beta o}^{\,3}} + \underline{G}(\lambda_\beta\, \underline{a}_{\beta o}) \right) \right]$$

$$\underline{t}_o^{(\underline{m},\, III)} / \underline{C}_o\, \underline{m}_o = \frac{2\,\underline{kT}}{3\eta\, \underline{e}\, (\underline{D}_\alpha \underline{Z}_\alpha - \underline{D}_\beta \underline{Z}_\beta)} \left[\underline{a}_{\alpha o}^{\,2} \sum_{\underline{n}=4}^{+\infty} \frac{1}{(\underline{n} - 2)} \, \frac{\underline{d}_{o\alpha,\underline{n}}}{\underline{kT}} \right.$$

$$\left. - \underline{a}_{\beta o}^{\,2} \sum_{\underline{n}=4}^{+\infty} \frac{1}{(\underline{n} - 2)} \, \frac{\underline{d}_{o\beta,\underline{n}}}{\underline{kT}} \right]$$

$$- \frac{2\pi}{\underline{e}} \frac{\underline{D}_o}{(\underline{D}_\alpha \underline{Z}_\alpha - \underline{D}_\beta \underline{Z}_\beta)} \left[\frac{\underline{D}_\alpha}{\underline{D}_o + \underline{D}_\alpha} \underline{a}_{\alpha o}^3 \sum_{\underline{n}=4}^{+\infty} \frac{1}{(\underline{n}+3)} \frac{\underline{d}_{o\alpha,\underline{n}}}{\underline{kT}} \right.$$

$$\left. - \frac{\underline{D}_\beta}{\underline{D}_o + \underline{D}_\beta} \underline{a}_{\beta o}^3 \sum_{\underline{n}=4}^{+\infty} \frac{1}{(\underline{n}+3)} \frac{\underline{d}_{o\beta,\underline{n}}}{\underline{kT}} \right] +$$

$$+ \frac{\underline{kT}}{3\eta\, \underline{e}\, (\underline{D}_\alpha \underline{Z}_\alpha - \underline{D}_\beta \underline{Z}_\beta)} \left[\underline{a}_{\alpha o}^2 \sum_{\underline{n},\underline{l}=4}^{+\infty} \frac{1}{(\underline{n}+\underline{l}-2)} \frac{\underline{d}_{o\alpha,\underline{n}}}{\underline{kT}} \frac{\underline{d}_{o\alpha,\underline{l}}}{\underline{kT}} \right.$$

$$\left. - \underline{a}_{\beta o}^2 \sum_{\underline{n},\underline{l}=4}^{+\infty} \frac{1}{(\underline{n}+\underline{l}-2)} \frac{\underline{d}_{o\beta,\underline{n}}}{\underline{kT}} \frac{\underline{d}_{o\beta,\underline{l}}}{\underline{kT}} \right]$$

$$+ \frac{2\pi}{3\underline{e}} \frac{\underline{D}_o}{(\underline{D}_\alpha \underline{Z}_\alpha - \underline{D}_\beta \underline{Z}_\beta)} \left[\frac{\underline{D}_\alpha}{\underline{D}_o + \underline{D}_\alpha} \underline{a}_{\alpha o}^3 \sum_{\underline{n},\underline{l}=4}^{+\infty} \frac{\underline{d}_{o\alpha,\underline{n}}}{\underline{kT}} \frac{\underline{d}_{o\alpha,\underline{l}}}{\underline{kT}} \left(\frac{1}{(\underline{n}+\underline{l})} \frac{\underline{B}_o^3}{\underline{a}_{\alpha o}^3} \right. \right.$$

$$\left. - \frac{\underline{n}^2(\underline{l}-3) + \underline{n}(\underline{l}^2 - 6\underline{l} - 3) - 6\underline{l}}{(\underline{n}+3)(\underline{n}+\underline{l})(\underline{n}+\underline{l}-3)} \right)$$

$$- \frac{\underline{D}_\beta}{\underline{D}_o + \underline{D}_\beta} \underline{a}_{\beta o}^3 \sum_{\underline{n},\underline{l}=4}^{+\infty} \frac{\underline{d}_{o\beta,\underline{n}}}{\underline{kT}} \frac{\underline{d}_{o\beta,\underline{l}}}{\underline{kT}} \left(\frac{1}{(\underline{n}+\underline{l})} \frac{\underline{B}_o^3}{\underline{a}_{\beta o}^3} \right.$$

$$\left. \left. - \frac{\underline{n}^2(\underline{l}-3) + \underline{n}(\underline{l}^2 - 6\underline{l} - 3) - 6\underline{l}}{(\underline{n}+3)(\underline{n}+\underline{l})(\underline{n}+\underline{l}-3)} \right) \right]$$

$$\underline{t}_o^{(\underline{m},\,IV)}/\underline{C}_o\,\underline{m}_o = -\frac{4\pi\epsilon\mu_o^2}{\underline{e}(2\epsilon+\epsilon_o)^2\,\underline{kT}}\;\frac{\underline{D}_o}{(\underline{D}_\alpha\underline{Z}_\alpha-\underline{D}_\beta\underline{Z}_\beta)}\left[\sum_{\underline{n}=4}^{+\infty}\frac{\underline{d}_{o\alpha,\underline{n}}}{\underline{kT}} - \sum_{\underline{n}=4}^{+\infty}\frac{\underline{d}_{o\beta,\underline{n}}}{\underline{kT}}\right].$$

When the only important short range forces are the electrostatic ion-cavity repulsive forces between the electrolyte ions and their image distributions in the spherical cavities of low dielectric constant created by the non-electrolyte molecules in the solvent, we can use for the coefficients $\underline{d}_{o\alpha,\underline{n}}$ and $\underline{d}_{o\beta,\underline{n}}$ the values given by equation D 13 and obtain the results:

(163)

$$\underline{t}_o^{(\underline{m},\,III)}/\underline{C}_o\,\underline{m}_o = -\frac{\underline{e}\,\underline{b}_o^3}{12\eta\epsilon\,(\underline{D}_\alpha\underline{Z}_\alpha-\underline{D}_\beta\underline{Z}_\beta)}\left[\frac{\underline{Z}_\alpha^2}{\underline{a}_{\alpha o}^2}\,\underline{H}(\underline{b}_o/\,\underline{a}_{\alpha o}) - \frac{\underline{Z}_\beta^2}{\underline{a}_{\beta o}^2}\,\underline{H}(\underline{b}_o/\,\underline{a}_{\beta o})\right]$$

$$+\frac{\pi\underline{e}\,\underline{b}_o^3}{14\epsilon\,\underline{kT}}\;\frac{\underline{D}_o}{(\underline{D}_\alpha\underline{Z}_\alpha-\underline{D}_\beta\underline{Z}_\beta)}\left[\frac{\underline{D}_\alpha}{\underline{D}_o+\underline{D}_\alpha}\,\frac{\underline{Z}_\alpha^2}{\underline{a}_{\alpha o}}\,\underline{K}(\underline{b}_o/\,\underline{a}_{\alpha o}) - \frac{\underline{D}_\beta}{\underline{D}_o+\underline{D}_\beta}\,\frac{\underline{Z}_\beta^2}{\underline{a}_{\beta o}}\,\underline{K}(\underline{b}_o/\,\underline{a}_{\beta o})\right]$$

$$t_o^{(m,\,IV)}/C_o\,m_o = \frac{\pi \mu_o^2\ e\ b_o^3}{(2\epsilon + \epsilon_o)^2\ (kT)^2}\ \frac{D_o}{(D_\alpha Z_\alpha - D_\beta Z_\beta)} \left[\frac{Z_\alpha^2}{a_{\alpha o}^4}\ L(b_o/\ a_{\alpha o}) - \frac{Z_\beta^2}{a_{\beta o}^4}\ L(b_o/\ a_{\beta o}) \right] .$$

Small higher order terms resulting from the short range forces have been neglected. The functions $H(x)$, $K(x)$, and $L(x)$ are defined by the series:

$$H(x) = \sum_{n=0}^{+\infty} \frac{2(\epsilon - \epsilon_o)}{\left[(n + 2)\epsilon + (n + 1)\,\epsilon_o\right]}\ x^{2n} \qquad (164)$$

$$K(x) = \sum_{n=0}^{+\infty} \frac{14\,(n + 1)(\epsilon - \epsilon_o)}{(2n + 7)\left[(n + 2)\epsilon + (n + 1)\,\epsilon_o\right]}\ x^{2n}$$

$$L(x) = \sum_{n=0}^{+\infty} \frac{2\,(n + 1)\,(\epsilon - \epsilon_o)}{\left[(n + 2)\epsilon + (n + 1)\,\epsilon_o\right]}\ x^{2n}$$

Each of these series is absolutely convergent for $0 \leq x < 1$.

Upon neglecting terms of order ϵ_o relative to terms of order ϵ (it is generally believed that the dielectric constant in the interior of organic molecules is about 2, whereas the dielectric constant of water is about 80); we may write the above series in closed form, namely,

(165)

$$\underline{H}(\underline{x}) = \sum_{\underline{n}=0}^{+\infty} \frac{2}{\underline{n}+2}\,\underline{x}^{2\underline{n}} = -\frac{2}{\underline{x}^2} - \frac{2}{\underline{x}^4}\ln(1+\underline{x})(1-\underline{x})$$

$$\underline{K}(\underline{x}) = \sum_{\underline{n}=0}^{+\infty} \frac{14(\underline{n}+1)}{(2\underline{n}+7)(\underline{n}+2)}\,\underline{x}^{2\underline{n}} =$$

$$= \frac{14}{3\underline{x}^4}\left[\left(1+\frac{5}{2\underline{x}^3}\right)\ln(1+\underline{x}) + \left(1-\frac{5}{2\underline{x}^3}\right)\ln(1-\underline{x}) - \frac{5}{\underline{x}^2} - \frac{5}{3}\right]$$

$$\underline{L}(\underline{x}) = \sum_{\underline{n}=0}^{+\infty} \frac{2(\underline{n}+1)}{\underline{n}+2}\,\underline{x}^{2\underline{n}} = \frac{2}{1-\underline{x}^2} + \frac{2}{\underline{x}^2} + \frac{2}{\underline{x}^4}\ln(1+\underline{x})(1-\underline{x}).$$

Values of these functions may be estimated from the tables in Appendix F.

For aqueous solutions at 0° C, with ϵ_o neglected relative to ϵ, with $\underline{t}_o^{(\underline{m})}/\underline{C}_o\,\underline{m}_o$ expressed in (cm^3/ Faraday), with μ_o in (Debye units), with $\underline{b}_o$, $\underline{a}_{\alpha o}$, and $\underline{a}_{\beta o}$ in (Angström units), with $\underline{D}_o$, $\underline{D}_\alpha$, and $\underline{D}_\beta$ in (units of 10^{-6} cm^2 / sec), and with $\underline{D}_o^{(rot)}$ in (units of 10^{10} sec^{-1}); equation 162 with $\underline{t}_o^{(\underline{m}, III)}/\underline{C}_o\,\underline{m}_o$ and $\underline{t}_o^{(\underline{m}, IV)}/\underline{C}_o\,\underline{m}_o$ given by equation 163 reduces to

(166)

$$\underline{t}_o^{(\underline{m}, I)}/\underline{C}_o\,\underline{m}_o = \frac{1.2616\,\underline{D}_o}{(\underline{D}_\alpha\underline{Z}_\alpha - \underline{D}_\beta\underline{Z}_\beta)}\left[\frac{\underline{D}_\alpha}{\underline{D}_o + \underline{D}_\alpha}\left(\underline{a}_{\alpha o}^3 - \underline{b}_o^3\right) + \right.$$

$$\left. - \frac{\underline{D}_\beta}{\underline{D}_o + \underline{D}_\beta}\left(\underline{a}_{\beta o}^3 - \underline{b}_o^3\right)\right]$$

$$\underline{t}_o^{(\underline{m}, II)}/\underline{C}_o\,\underline{m}_o = \frac{33.478\,\mu_o^2}{(\underline{D}_\alpha\underline{Z}_\alpha - \underline{D}_\beta\underline{Z}_\beta)}\left[\frac{\underline{Z}_\alpha^2}{\underline{a}_{\alpha o}^2} - \frac{\underline{Z}_\beta^2}{\underline{a}_{\beta o}^2}\right] +$$

$$+ \frac{0.99694\,\mu_o^2\,\underline{D}_o}{(\underline{D}_\alpha\underline{Z}_\alpha - \underline{D}_\beta\underline{Z}_\beta)}\left[\frac{\underline{D}_\alpha}{\underline{D}_o + \underline{D}_\alpha}\frac{\underline{Z}_\alpha^2}{\underline{a}_{\alpha o}}\left(\frac{\underline{b}_o^3}{\underline{a}_{\alpha o}^3} + \underline{G}(\lambda_\alpha\,\underline{a}_{\alpha o})\right) + \right.$$

$$\left. - \frac{\underline{D}_\beta}{\underline{D}_o + \underline{D}_\beta}\frac{\underline{Z}_\beta^2}{\underline{a}_{\beta o}}\left(\frac{\underline{b}_o^3}{\underline{a}_{\beta o}^3} + \underline{G}(\lambda_\beta\,\underline{a}_{\beta o})\right)\right]$$

$$\underline{t}_o^{(\underline{m}, III)}/\underline{C}_o\,\underline{m}_o = - \frac{73.832\,\underline{b}_o^3}{(\underline{D}_\alpha\underline{Z}_\alpha - \underline{D}_\beta\underline{Z}_\beta)}\left[\frac{\underline{Z}_\alpha^2}{\underline{a}_{\alpha o}^2}\,\underline{H}(\underline{b}_o/\underline{a}_{\alpha o}) \right.$$

$$\left. - \frac{\underline{Z}_\beta^2}{\underline{a}_{\beta o}^2}\,\underline{H}(\underline{b}_o/\underline{a}_{\beta o})\right] +$$

$$+ \frac{0.94235\, \underline{b}_o^3 \underline{D}_o}{(\underline{D}_\alpha \underline{Z}_\alpha - \underline{D}_\beta \underline{Z}_\beta)} \left[\frac{\underline{D}_\alpha}{\underline{D}_o + \underline{D}_\alpha} \frac{\underline{Z}_\alpha^2}{\underline{a}_{\alpha o}} \underline{K}(\underline{b}_o / \underline{a}_{\alpha o}) - \frac{\underline{D}_\beta}{\underline{D}_o + \underline{D}_\beta} \frac{\underline{Z}_\beta^2}{\underline{a}_{\beta o}} \underline{K}(\underline{b}_o / \underline{a}_{\beta o}) \right]$$

$$\underline{t}_o^{(\underline{m},\, IV)} / \underline{C}_o \underline{m}_o = \frac{0.99694\, \mu_o^2 \underline{b}_o^3 \underline{D}_o}{(\underline{D}_\alpha \underline{Z}_\alpha - \underline{D}_\beta \underline{Z}_\beta)} \left[\frac{\underline{Z}_\alpha^2}{\underline{a}_{\alpha o}^4} \underline{L}(\underline{b}_o / \underline{a}_{\alpha o}) - \frac{\underline{Z}_\beta^2}{\underline{a}_{\beta o}^4} \underline{L}(\underline{b}_o / \underline{a}_{\beta o}) \right] .$$

The numerical constants appearing in these equations are based on the values of the physical constants tabulated in Appendices G and H. Similarly, for aqueous solutions at 25° C, we have

(167)

$$\underline{t}_o^{(\underline{m},\, I)} / \underline{C}_o \underline{m}_o = \frac{1.2616\, \underline{D}_o}{(\underline{D}_\alpha \underline{Z}_\alpha - \underline{D}_\beta \underline{Z}_\beta)} \left[\frac{\underline{D}_\alpha}{\underline{D}_o + \underline{D}_\alpha} \left(\underline{a}_{\alpha o}^3 - \underline{b}_o^3 \right) - \frac{\underline{D}_\beta}{\underline{D}_o + \underline{D}_\beta} \left(\underline{a}_{\beta o}^3 - \underline{b}_o^3 \right) \right] +$$

$$\underline{t}_o^{(\underline{m},\, II)} / \underline{C}_o \underline{m}_o = \frac{77.298\, \mu_o^2}{(\underline{D}_\alpha \underline{Z}_\alpha - \underline{D}_\beta \underline{Z}_\beta)} \left[\frac{\underline{Z}_\alpha^2}{\underline{a}_{\alpha o}^2} - \frac{\underline{Z}_\beta^2}{\underline{a}_{\beta o}^2} \right] +$$

$$+ \frac{1.0506\,\mu_o^2\,\underline{D}_o}{(\underline{D}_\alpha \underline{Z}_\alpha - \underline{D}_\beta \underline{Z}_\beta)} \left[\frac{\underline{D}_\alpha}{\underline{D}_o + \underline{D}_\alpha} \frac{\underline{Z}_\alpha^2}{\underline{a}_{\alpha o}} \left(\frac{\underline{b}_o^3}{\underline{a}_{\alpha o}^3} + \underline{G}(\lambda_\alpha \underline{a}_{\alpha o}) \right) \right. +$$

$$\left. - \frac{\underline{D}_\beta}{\underline{D}_o + \underline{D}_\beta} \frac{\underline{Z}_\beta^2}{\underline{a}_{\beta o}} \left(\frac{\underline{b}_o^3}{\underline{a}_{\beta o}^3} + \underline{G}(\lambda_\beta \underline{a}_{\beta o}) \right) \right]$$

$$\underline{t}_o^{(\underline{m},\,III)} / \underline{C}_o \underline{m}_o = - \frac{166.05\,\underline{b}_o^3}{(\underline{D}_\alpha \underline{Z}_\alpha - \underline{D}_\beta \underline{Z}_\beta)} \left[\frac{\underline{Z}_\alpha^2}{\underline{a}_{\alpha o}^2} \underline{H}(\underline{b}_o / \underline{a}_{\alpha o}) \right.$$

$$\left. - \frac{\underline{Z}_\beta^2}{\underline{a}_{\beta o}^2} \underline{H}(\underline{b}_o / \underline{a}_{\beta o}) \right] +$$

$$+ \frac{0.96734\,\underline{b}_o^3\,\underline{D}_o}{(\underline{D}_\alpha \underline{Z}_\alpha - \underline{D}_\beta \underline{Z}_\beta)} \left[\frac{\underline{D}_\alpha}{\underline{D}_o + \underline{D}_\alpha} \frac{\underline{Z}_\alpha^2}{\underline{a}_{\alpha o}} \underline{K}(\underline{b}_o / \underline{a}_{\alpha o}) \right.$$

$$\left. - \frac{\underline{D}_\beta}{\underline{D}_o + \underline{D}_\beta} \frac{\underline{Z}_\beta^2}{\underline{a}_{\beta o}} \underline{K}(\underline{b}_o / \underline{a}_{\beta o}) \right]$$

$$\underline{t}_o^{(\underline{m},\,IV)} / \underline{C}_o \underline{m}_o = \frac{1.0506\,\mu_o^2\,\underline{b}_o^3\,\underline{D}_o}{(\underline{D}_\alpha \underline{Z}_\alpha - \underline{D}_\beta \underline{Z}_\beta)} \left[\frac{\underline{Z}_\alpha^2}{\underline{a}_{\alpha o}^4} \underline{L}(\underline{b}_o / \underline{a}_{\alpha o}) \right.$$

$$\left. - \frac{\underline{Z}_\beta^2}{\underline{a}_{\beta o}^4} \underline{L}(\underline{b}_o / \underline{a}_{\beta o}) \right] .$$

Tables IV and V summarize the computation of theoretical mass transport ratios, using equations 166 and 167, for the systems which have been studied experimentally. Values of the ionic and molecular sizes, the dipole moments, and the diffusion coefficients that were used in these calculations are given in Appendices I and J. For the closest approach parameters, $\underline{a}_{\alpha o}$ and $\underline{a}_{\beta o}$, sums of the ionic and molecular radii, $\underline{b}_o + \underline{b}_\alpha$ and $\underline{b}_o + \underline{b}_\beta$, were used.

Comparison of Theory with Experiment

Agreement between theory and experiment is plausible; in some cases it is extremely good, but this may be fortuitous.

One difficulty is the lack of sufficient experimental data so as to provide a real test of the theory. The systems on which transport measurements have been made are not necessarily those which one would expect to be most amenable to representation by our model. Furthermore, the available data cannot be accurately extrapolated to infinite dilution. Data obtained by the Hittorf method is not very reliable; and in calculating mass transport ratios from this data we have not made allowances for the possible transport of water of hydration by the electrolyte.

Another complication is ion binding, i.e., the associa-

Table IV

Theoretical Non-electrolyte Transport, at 0° C.

Non-electrolyte	Electrolyte	$\frac{t_o(m, I)}{C_o m_o}$	$\frac{t_o(m, II)}{C_o m_o}$ Term 1	Term 2	$\frac{t_o(m, III)}{C_o m_o}$ Term 1	Term 2	$\frac{t_o(m, IV)}{C_o m_o}$	$\frac{t_o(m)}{C_o m_o}$
Urea	LiCl	-19.54	+2.10	+0.51	-10.40	+1.05	+5.18	-21.1
Urea	NaCl	-13.86	+1.13	+0.33	- 4.68	+0.53	+1.68	-14.9
Resorcinol	LiCl	-15.82	+0.36	+0.07	-12.18	+1.21	+0.70	-25.7
Mannitol	LiCl	-18.54	+0.60	+0.12	-20.17	+2.54	+1.89	-33.6
Raffinose	LiCl	-20.01	+0.21	+0.04	-27.33	+3.79	+0.88	-42.4
Raffinose	NaCl	-13.52	+0.12	+0.03	-12.89	+1.82	+0.33	-24.1
Raffinose	KCl	- 6.21	+0.05	+0.03	- 4.69	+0.77	+0.10	-10.0
Raffinose	NH_4Cl	- 4.43	+0.04	+0.02	- 2.97	+0.47	+0.06	- 6.8
Sucrose	NaF	- 6.29	+0.09	+0.04	- 7.79	+1.18	+0.30	-12.5
Sucrose	NaCl	-13.74	+0.15	+0.04	-11.23	+1.54	+0.38	-22.9
Sucrose	NaI	-20.30	+0.20	+0.06	-13.72	+1.96	+0.43	-31.4

Table V

Theoretical Non-electrolyte Transport, at 25° C.

Non-electrolyte	Electrolyte	$\frac{t_o(m, I)}{C_o m_o}$	$\frac{t_o(m, II)}{C_o m_o}$		$\frac{t_o(m, III)}{C_o m_o}$		$\frac{t_o(m, IV)}{C_o m_o}$	$\frac{t_o(m)}{C_o m_o}$
			Term 1	Term 2	Term 1	Term 2		
Raffinose	LiCl	-21.31	+0.24	+0.05	-29.57	+4.21	+1.01	-45.4
Raffinose	NaCl	-14.71	+0.14	+0.04	-14.18	+2.00	+0.38	-26.3
Raffinose	KCl	- 6.97	+0.06	+0.03	- 5.26	+0.86	+0.12	-11.2
Raffinose	CaCl	- 2.18	+0.02	+0.01	- 1.40	+0.25	+0.03	- 3.3
Urea	NaCl	-13.96	+1.28	+0.36	- 4.14	+0.56	+1.81	-15.1
Maltose	$(CH_3)_4NI$	+24.86	-0.14	-0.12	+ 5.97	-1.18	-0.09	+29.3
Maltose	$(C_2H_5)_4NI$	+36.15	-0.21	-0.17	+ 8.23	-1.71	-0.12	+42.2
Maltose	$(C_3H_7)_4NI$	+45.30	-0.27	-0.21	+10.32	-2.23	-0.14	+52.8
Acetone	$(CH_3)_4NI$	+24.61	-0.27	-0.21	+ 2.86	-0.86	-0.15	+26.0
Ethyl Acetate	$(CH_3)_4NI$	+22.93	-0.11	-0.08	+ 2.45	-0.69	-0.05	+24.5
Ethyl Acetate	$(C_2H_5)_4NI$	+30.90	-0.16	-0.10	+ 3.33	-0.98	-0.07	+32.9
Ethyl Acetate	$(C_4H_9)_4NI$	+40.74	-0.23	-0.12	+ 4.63	-1.41	-0.08	+43.5
Glycine	$BaCl_2$	-10.16	+91.32	+8.44	-27.50	+3.35	+40.35	+105.8

tion of an ion with a neutral non-electrolyte molecule to form a charged complex. It may well be that binding occurs between the iodide ions and the non-electrolytes: maltose, acetone, and ethyl acetate. If so, this could account for the rather large observed transport of these non-electrolytes in the direction opposite to that predicted by our theory.

In Figure 2 values of $-\underline{t}_0^{(\underline{m})}/\underline{C}_0\,\underline{m}_0$ from Table II in which raffinose is the non-electrolyte are plotted as ordinate against the square root of the electrolyte concentration as abscissa. With this abscissa scale the three points for sodium chloride, and also those for lithium chloride, fall on a straight line. The shaded points in the figure are the theoretical values taken from Table IV. For lithium chloride, and even for sodium chloride, the agreement between theory and experiment is quite satisfactory.

In Figure 3 values of $+\underline{t}_0^{(\underline{m})}/\underline{C}_0\,\underline{m}_0$ from Table III in which glycine is the non-electrolyte are plotted against the square root of the barium chloride concentration. The shaded point is the theoretical value taken from Table V.

One tentative conclusion drawn from these studies is that: for the systems considered here the observed non-electrolyte transport cannot be accounted for in terms of a simple model whereby the solute particles are charged rigid spheres interacting with one another solely through the long range ion-ion and ion-dipole electrostatic forces.

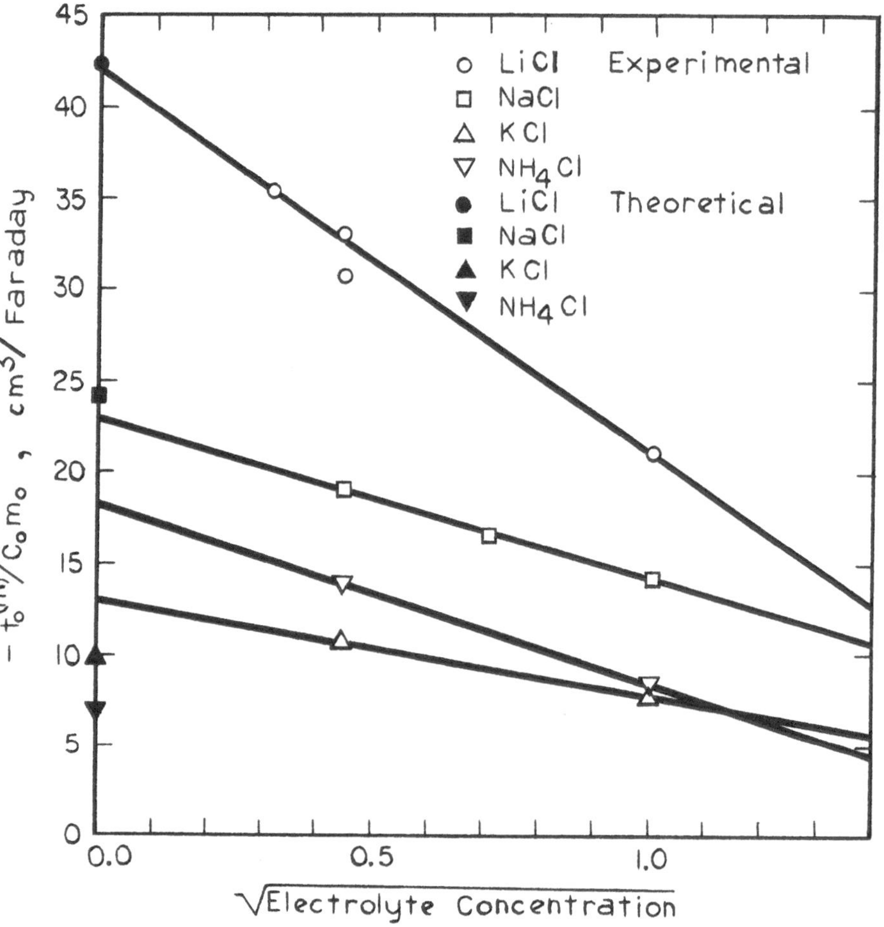

Figure 2. The Variation of $-\underline{t}_o^{(\underline{m})}/\underline{C}_o\underline{m}_o$ in 1% to 2% Raffinose Solutions with the Nature and Concentration of the Added Electrolyte.

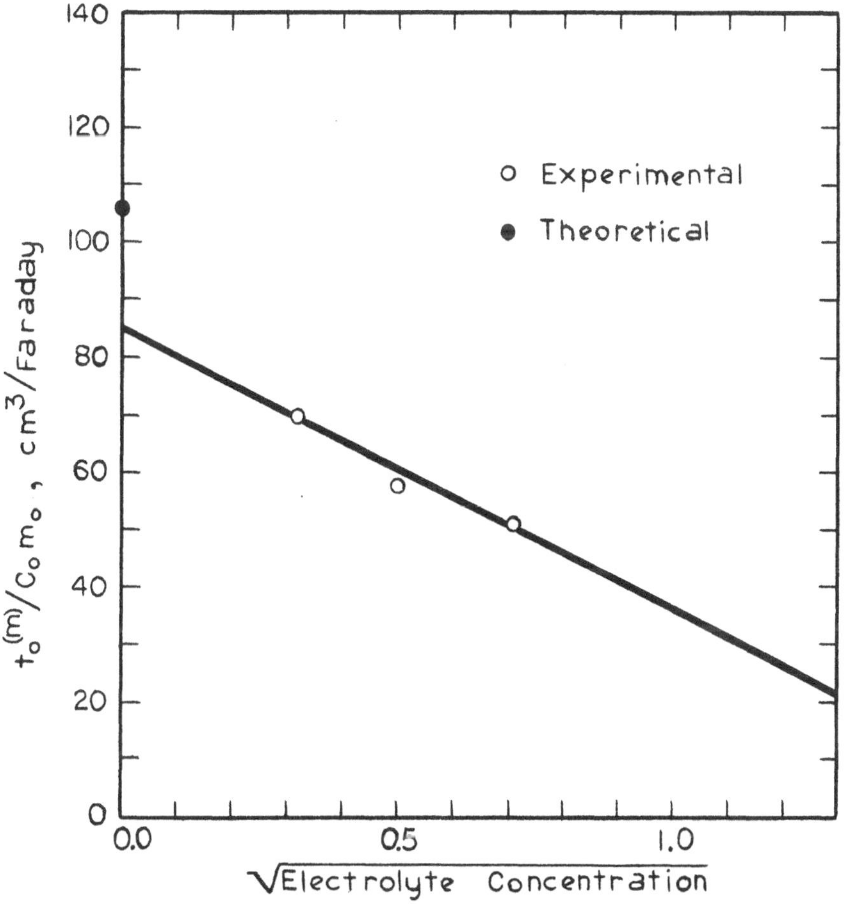

Figure 3. The Variation of $\underline{t}_o^{(\underline{m})}/\underline{C}_o\,\underline{m}_o$ in 1 M Glycine Solutions with the Concentration of the Added Electrolyte, $BaCl_2$.

Short range forces, perhaps the electrostatic ion-cavity repulsive forces, appear to be important.

Final judgement as to the importance of the short range forces, indeed, final judgement on the validity of this theory of the electrolytic transport of non-electrolytes for infinitely dilute solutions must await further measurements, at high dilution, by the moving boundary method. At that time, a decision can be made as to whether or not it might be worthwhile to consider the extension of the theory to higher electrolyte concentrations (neglecting short range forces) as given in the author's dissertation.[45]

Acknowledgments

I wish to acknowledge a number of very helpful discussions with the late Professor John G. Kirkwood and with Professor Lars Onsager, who jointly directed the dissertation on which this paper is based.

Appendix A: Surface Harmonics

We may associate with the spherical polar coordinates (r, θ, φ) expressions of the form

$$S_n(\theta, \varphi) = A_n P_n(\cos \theta) + \sum_{m=1}^{n} \left(A_n^{(m)} \cos m\varphi + B_n^{(m)} \sin m\varphi \right) P_n^{|m|}(\cos \theta) \qquad (A\,1)$$

where n is a positive integer; where A_n, $A_n^{(m)}$, and $B_n^{(m)}$ are arbitrary constants; where P_n is the Legendre function of degree n; and where P_n^{m} is the associated Legendre function of degree n and order m. This expression is called a surface harmonic of degree n on the sphere r = constant.

By employing a vector notation, these surface harmonics can be written in a more concise form. For example:

$$S_0(\theta, \varphi) = c = c\ P_0(\cos \theta) \qquad (A\,2)$$

$$S_1(\theta, \varphi) = \tilde{c} \cdot \tilde{1}_r = c_x P_1^1(\cos \theta) \cos \varphi + c_y P_1^1(\cos \theta) \sin \varphi$$

$$+ \; c_z \, P_1(\cos\theta)$$

$$S_2(\theta,\varphi) = \mathbf{c}\cdot\mathbf{1}_r \; \mathbf{c}'\cdot\mathbf{1}_r - \tfrac{1}{3}\,\mathbf{c}\cdot\mathbf{c}' =$$

$$= \tfrac{1}{3}\,(2\,c_z c_z' - c_x c_x' - c_y c_y')\; P_2(\cos\theta) \; +$$

$$+ \; \tfrac{1}{3}\,(c_x c_z' + c_x' c_z)\; P_2^1(\cos\theta)\cos\varphi \; +$$

$$+ \; \tfrac{1}{3}\,(c_y c_z' + c_z c_y')\; P_2^1(\cos\theta)\sin\varphi \; +$$

$$+ \; \tfrac{1}{6}\,(c_x c_x' - c_y c_y')\; P_2^2(\cos\theta)\cos 2\varphi \; +$$

$$+ \; \tfrac{1}{6}\,(c_x c_y' + c_y c_x')\; P_2^2(\cos\theta)\sin 2\varphi$$

where $\mathbf{1}_r$, $\mathbf{1}_\theta$, and $\mathbf{1}_\varphi$ are the orthonormal basic vectors associated with the spherical polar coordinates (r, θ, φ); and where c, $\mathbf{c} = c_x\,\mathbf{i} + c_y\,\mathbf{j} + c_z\,\mathbf{k}$, and $\mathbf{c}' = c_x'\,\mathbf{i} + c_y'\,\mathbf{j} + c_z'\,\mathbf{k}$ are constants.

The surface harmonics $S_n(\theta,\varphi)$ form a complete orthogonal system of functions on the sphere r = constant; and every function $g(\theta,\varphi)$ which is continuous together with its derivatives up to the second order on the sphere may be expanded in an absolutely and uniformly convergent series in terms of them. Hence

$$g(\theta,\varphi) = \sum_{n=0}^{+\infty} S_n(\theta,\varphi) \qquad \text{(A 3)}$$

where

$$A_n = \frac{2n+1}{4\pi}\int_0^{2\pi}\int_0^{\pi} g(\theta,\varphi)\, P_n(\cos\theta)\sin\theta\, d\theta\, d\varphi$$

$$A_n^{(m)} = \frac{2n+1}{2\pi}\,\frac{(n-|m|)!}{(n+|m|)!}\int_0^{2\pi}\int_0^{\pi} g(\theta,\varphi)\, P_n^{|m|}(\cos\theta)\cos m\varphi\,\sin\theta\, d\theta\, d\varphi$$

$$B_n^{(m)} = \frac{2n+1}{2\pi}\,\frac{(n-|m|)!}{(n+|m|)!}\int_0^{2\pi}\int_0^{\pi} g(\theta,\varphi)\, P_n^{|m|}(\cos\theta)\sin m\varphi\,\sin\theta\, d\theta\, d\varphi.$$

Appendix B: Selected Differential Properties of the Surface Harmonics

Upon operating on the surface harmonics $\mathbf{S}_n(\theta, \varphi)$ with the gradient operator

$$\nabla_{\mathbf{r}} = \mathbf{1}_r \frac{\partial}{\partial r} + \mathbf{1}_\theta \frac{1}{r} \frac{\partial}{\partial \theta} + \mathbf{1}_\varphi \frac{1}{r \sin\theta} \frac{\partial}{\partial \varphi} \tag{B 1}$$

associated with the spherical polar coordinates $\mathbf{r} = (r, \theta, \varphi)$, we find that

$$\nabla_{\mathbf{r}}\, \mathbf{c} = 0 \tag{B 2}$$

$$\nabla_{\mathbf{r}} (\mathbf{c} \cdot \mathbf{1}_r) = \frac{1}{r} (\mathbf{c} - \mathbf{c} \cdot \mathbf{1}_r \mathbf{1}_r)$$

$$\nabla_{\mathbf{r}} (\mathbf{c} \cdot \mathbf{1}_r\; \mathbf{c}' \cdot \mathbf{1}_r - \tfrac{1}{3} \mathbf{c} \cdot \mathbf{c}') = \frac{1}{r}(\mathbf{c}\,\mathbf{c}' + \mathbf{c}'\,\mathbf{c}) \cdot \mathbf{1}_r - \frac{2}{r}\, \mathbf{c} \cdot \mathbf{1}_r\; \mathbf{c}' \cdot \mathbf{1}_r\; \mathbf{1}_r .$$

Upon operating on the surface harmonics $\mathbf{S}_n(\theta, \varphi)$ with the Laplacian operator $\nabla_{\mathbf{r}}^2$, defined by

$$\nabla_{\mathbf{r}}^2 f = \nabla_{\mathbf{r}} \cdot \nabla_{\mathbf{r}} f = \frac{1}{r^2}\frac{\partial}{\partial r}\left(r^2 \frac{\partial f}{\partial r}\right) + \frac{1}{r^2 \sin\theta}\frac{\partial}{\partial \theta}\left(\sin\theta \frac{\partial f}{\partial \theta}\right) + \frac{1}{r^2 \sin^2\theta}\frac{\partial^2 f}{\partial \varphi^2} \tag{B 3}$$

where f is any scalar field, we have

$$\nabla_{\mathbf{r}}^2 \mathbf{c} = 0 \tag{B 4}$$

$$\nabla_{\mathbf{r}}^2 (\mathbf{c} \cdot \mathbf{1}_r) = -\frac{2}{r^2}\, \mathbf{c} \cdot \mathbf{1}_r$$

$$\nabla_{\mathbf{r}}^2 (\mathbf{c} \cdot \mathbf{1}_r\, \mathbf{c}' \cdot \mathbf{1}_r - \tfrac{1}{3}\, \mathbf{c} \cdot \mathbf{c}') = -\frac{6}{r^2}(\mathbf{c} \cdot \mathbf{1}_r\, \mathbf{c}' \cdot \mathbf{1}_r - \tfrac{1}{3}\, \mathbf{c} \cdot \mathbf{c}') .$$

The surface harmonic $S_n(\theta, \varphi)$ of degree n multiplied by r^n or by r^{-n-1} is called a spherical harmonic of degree n. It is a solution of Laplace's equation, i.e.,

$$\nabla_{\underset{\sim}{\underline{r}}}^2 \left(\underline{r}^{\underline{n}} \ \underline{S}_{\underline{n}}(\theta, \varphi) \right) = 0 \qquad (B5)$$

$$\nabla_{\underset{\sim}{\underline{r}}}^2 \left(\underline{r}^{-\underline{n} - 1} \ \underline{S}_{\underline{n}}(\theta, \varphi) \right) = 0 .$$

The general solution of Laplace's equation is a linear combination of these spherical harmonics.

Operating on the surface harmonics $\underline{S}_{\underline{n}}(\Theta, \Phi)$ with the "rotational" gradient operator

$$\nabla_{\underset{\sim}{\underline{R}}} = \underset{\sim}{1}_R \frac{\partial}{\partial \Xi} + \underset{\sim}{1}_\Phi \frac{\partial}{\partial \Theta} + \underset{\sim}{1}_\Theta \left(- \frac{1}{\sin \Theta} \frac{\partial}{\partial \Phi} + \frac{\cos \Theta}{\sin \Theta} \frac{\partial}{\partial \Xi} \right) \qquad (B6)$$

associated with the Eulerian angles $\underset{\sim}{\underline{R}} = (\Phi, \Theta, \Xi)$, we obtain the results

$$\nabla_{\underset{\sim}{\underline{R}}} \ \underline{c} = 0 \qquad (B7)$$

$$\nabla_{\underset{\sim}{\underline{R}}} (\underset{\sim}{\underline{c}} \cdot \underset{\sim}{1}_{\underline{R}}) = \underset{\sim}{1}_{\underline{R}} \times \underset{\sim}{\underline{c}}$$

$$\nabla_{\underset{\sim}{\underline{R}}} (\underset{\sim}{\underline{c}} \cdot \underset{\sim}{1}_{\underline{R}} \ \underset{\sim}{\underline{c}}' \cdot \underset{\sim}{1}_{\underline{R}} - \tfrac{1}{3} \underset{\sim}{\underline{c}} \cdot \underset{\sim}{\underline{c}}') = \underset{\sim}{\underline{c}} \cdot \underset{\sim}{1}_{\underline{R}} \ \underset{\sim}{1}_{\underline{R}} \times \underset{\sim}{\underline{c}}'$$

$$+ \ \underset{\sim}{\underline{c}}' \cdot \underset{\sim}{1}_{\underline{R}} \ \underset{\sim}{1}_{\underline{R}} \times \underset{\sim}{\underline{c}} .$$

Upon operating on the surface harmonics $\underline{S}_{\underline{n}}(\Theta, \Phi)$ with the "rotational" Laplacian operator $\nabla_{\underline{R}}^2$, defined by

$$\nabla_{\underline{R}}^2 \underline{f} = \nabla_{\underline{R}} \cdot \nabla_{\underline{R}} \underline{f} = + \frac{1}{\sin^2 \Theta} \frac{\partial^2 \underline{f}}{\partial \Xi^2} + \frac{1}{\sin^2 \Theta} \frac{\partial^2 \underline{f}}{\partial \Phi^2} + \frac{1}{\sin \Theta} \frac{\partial}{\partial \Theta} \left(\sin \Theta \frac{\partial \underline{f}}{\partial \Theta} \right) - \frac{2 \cos \Theta}{\sin^2 \Theta} \frac{\partial^2 \underline{f}}{\partial \Phi \, \partial \Xi} \tag{B8}$$

where $\underline{f}$ is any scalar field, we have

$$\nabla_{\underline{R}}^2 \underline{c} = 0 \tag{B9}$$

$$\nabla_{\underline{R}}^2 (\underline{c} \cdot \underline{1}_{\underline{R}}) = -2 \, \underline{c} \cdot \underline{1}_{\underline{R}}$$

$$\nabla_{\underline{R}}^2 (\underline{c} \cdot \underline{1}_{\underline{R}} \, \underline{c}' \cdot \underline{1}_{\underline{R}} - \tfrac{1}{3} \underline{c} \cdot \underline{c}') = -6 (\underline{c} \cdot \underline{1}_{\underline{R}} \, \underline{c}' \cdot \underline{1}_{\underline{R}} - \tfrac{1}{3} \underline{c} \cdot \underline{c}').$$

Appendix C: Calculation of Integrals over Angles of Vectors and Tensors

Consider the integral

$$\int_0^{2\pi}\int_0^{\pi} \underline{f}(\theta,\varphi) \sin\theta \, d\theta \, d\varphi \tag{C 1}$$

in which the integrand $\underline{f}(\theta,\varphi)$ may be a scalar, a vector, or a tensor. Frequently, integrals of this type can be most readily evaluated by expanding the integrand in terms of the orthonormal basic vectors $\underline{1}_r$, $\underline{1}_\theta$, and $\underline{1}_\varphi$ associated with the spherical polar coordinates (r, θ, φ); and by utilizing the following table of integrals:

$$\int_0^{2\pi}\int_0^{\pi} \underline{c} \sin\theta \, d\theta \, d\varphi = 4\pi \underline{c} \tag{C 2}$$

$$\int_0^{2\pi}\int_0^{\pi} \underline{1}_r \sin\theta \, d\theta \, d\varphi = 0$$

$$\int_0^{2\pi}\int_0^{\pi} \underline{1}_\theta \sin\theta \, d\theta \, d\varphi = -\pi^2 \underline{k}$$

$$\int_0^{2\pi}\int_0^{\pi} \mathbf{1}_{\varphi} \sin\theta \, d\theta \, d\varphi = 0$$

$$\int_0^{2\pi}\int_0^{\pi} \mathbf{1}_{r}\mathbf{1}_{r} \sin\theta \, d\theta \, d\varphi = \frac{4\pi}{3} \mathbf{U}$$

$$\int_0^{2\pi}\int_0^{\pi} \mathbf{1}_{\theta}\mathbf{1}_{\theta} \sin\theta \, d\theta \, d\varphi = \frac{2\pi}{3} \mathbf{U} + 2\pi \mathbf{k}\mathbf{k}$$

$$\int_0^{2\pi}\int_0^{\pi} \mathbf{1}_{\varphi}\mathbf{1}_{\varphi} \sin\theta \, d\theta \, d\varphi = 2\pi \mathbf{U} - 2\pi \mathbf{k}\mathbf{k}$$

$$\int_0^{2\pi}\int_0^{\pi} \mathbf{1}_{r}\mathbf{1}_{r}\mathbf{1}_{r} \sin\theta \, d\theta \, d\varphi = 0$$

$$\int_0^{2\pi}\int_0^{\pi} \mathbf{1}_{r}\mathbf{1}_{r}\mathbf{1}_{r}\mathbf{1}_{r} \sin\theta \, d\theta \, d\varphi = \frac{4\pi}{15} \Big(\mathbf{U}\mathbf{U} + [\![\mathbf{U}]\!] + [\![\mathbf{I}]\!] \Big)$$

in which: c is a constant; $\mathbf{U}$ is the unit dyadic defined by

$$\underset{\sim}{U} = \underset{\sim}{i}\underset{\sim}{i} + \underset{\sim}{j}\underset{\sim}{j} + \underset{\sim}{k}\underset{\sim}{k} \; ; \qquad (C\,3)$$

$[\![\underset{\sim}{U}]\!]$ is the unit tetradic defined by

$$\begin{aligned}[\![\underset{\sim}{U}]\!] = \; & \underset{\sim}{i}\underset{\sim}{i}\underset{\sim}{i}\underset{\sim}{i} + \underset{\sim}{j}\underset{\sim}{j}\underset{\sim}{j}\underset{\sim}{j} + \underset{\sim}{k}\underset{\sim}{k}\underset{\sim}{k}\underset{\sim}{k} + \\ & + \underset{\sim}{j}\underset{\sim}{i}\underset{\sim}{i}\underset{\sim}{j} + \underset{\sim}{i}\underset{\sim}{j}\underset{\sim}{j}\underset{\sim}{i} + \underset{\sim}{i}\underset{\sim}{k}\underset{\sim}{k}\underset{\sim}{i} + \\ & + \underset{\sim}{k}\underset{\sim}{i}\underset{\sim}{i}\underset{\sim}{k} + \underset{\sim}{k}\underset{\sim}{j}\underset{\sim}{j}\underset{\sim}{k} + \underset{\sim}{j}\underset{\sim}{k}\underset{\sim}{k}\underset{\sim}{j} \; ;\end{aligned} \qquad (C\,4)$$

and $[\![\underset{\sim}{I}]\!]$ is the "inversion" tetradic defined by

$$\begin{aligned}[\![\underset{\sim}{I}]\!] = \; & \underset{\sim}{i}\underset{\sim}{i}\underset{\sim}{i}\underset{\sim}{i} + \underset{\sim}{j}\underset{\sim}{j}\underset{\sim}{j}\underset{\sim}{j} + \underset{\sim}{k}\underset{\sim}{k}\underset{\sim}{k}\underset{\sim}{k} + \\ & + \underset{\sim}{i}\underset{\sim}{j}\underset{\sim}{i}\underset{\sim}{j} + \underset{\sim}{j}\underset{\sim}{i}\underset{\sim}{j}\underset{\sim}{i} + \underset{\sim}{i}\underset{\sim}{k}\underset{\sim}{i}\underset{\sim}{k} + \\ & + \underset{\sim}{k}\underset{\sim}{i}\underset{\sim}{k}\underset{\sim}{i} + \underset{\sim}{j}\underset{\sim}{k}\underset{\sim}{j}\underset{\sim}{k} + \underset{\sim}{k}\underset{\sim}{j}\underset{\sim}{k}\underset{\sim}{j} \; .\end{aligned} \qquad (C\,5)$$

Appendix D: Some Electrostatic Problems

We shall present here solutions to some of the important electrostatic problems suggested by the model which we have chosen to represent a solution containing non-electrolyte molecules and electrolyte ions.

The Pair Potential Energy of a Non-electrolyte Molecule and an Electrolyte Ion

The mutual potential energy of a non-electrolyte molecule of species o in configuration $\underset{\sim}{q}_o$ and an electrolyte ion of species α in configuration $\underset{\sim}{q}_\alpha$ is denoted by $\underline{V}_{o\alpha}(\underset{\sim}{q}_o, \underset{\sim}{q}_\alpha)$. This represents the work required to bring a non-electrolyte molecule of species o and an electrolyte ion of species α from infinite separation in the pure solvent to the given configuration. The electrostatic work required to perform this operation is equal to the electrostatic work of charging the system in the given configuration minus the electrostatic work of charging when the non-electrolyte molecule and the electrolyte ion are infinitely separated.[46] Before we can use these charging processes to calculate the electrostatic contribution to $\underline{V}_{o\alpha}(\underset{\sim}{q}_o, \underset{\sim}{q}_\alpha)$ we must determine: (1) the potential at any point in the electrostatic field of a non-electrolyte molecule, (2) the potential at any point in the electrostatic field of an electrolyte ion, and (3) the potential at any point in the electrostatic field of both a non-electrolyte molecule and an electrolyte ion.

The potential $\Psi_o(\underline{r}_o + \underline{r} \mid \underline{q}_o)$ at the point $\underline{r}_o + \underline{r}$ due to a non-electrolyte molecule of species o at $\underline{q}_o$ satisfies Poisson's equation

$$\nabla_{\underline{r}}^2 \quad \Psi_o^{(\underline{r} > \underline{b}_o)}(\underline{r}_o + \underline{r} \mid \underline{q}_o) = 0 \qquad \text{(D 1)}$$

$$\nabla_{\underline{r}}^2 \quad \Psi_o^{(\underline{r} < \underline{b}_o)}(\underline{r}_o + \underline{r} \mid \underline{q}_o) = -\frac{4\pi}{\epsilon_o} \sum_{\underline{i}=1}^{\underline{s}_o} \underline{e}_{\underline{i}} \, \delta(\underline{r} - \underline{r}_{\underline{i}})$$

as well as the boundary conditions

$$\Psi_o^{(\underline{r} < \underline{b}_o)}(\underline{r}_o + \underline{r} \mid \underline{q}_o) = \Psi_o^{(\underline{r} > \underline{b}_o)}(\underline{r}_o + \underline{r} \mid \underline{q}_o) \qquad \text{(D 2)}$$

$$\underline{1}_{\underline{r}} \times \nabla_{\underline{r}} \quad \Psi_o^{(\underline{r} < \underline{b}_o)}(\underline{r}_o + \underline{r} \mid \underline{q}_o) =$$

$$= \underline{1}_{\underline{r}} \times \nabla_{\underline{r}} \quad \Psi_o^{(\underline{r} > \underline{b}_o)}(\underline{r}_o + \underline{r} \mid \underline{q}_o)$$

$$\epsilon_0 \, \mathbf{1}_r \cdot \nabla_{\mathbf{r}} \; \Psi_0^{(r < b_0)}(\mathbf{r}_0 + \mathbf{r} \mid \mathbf{q}_0) =$$

$$= \epsilon \, \mathbf{1}_r \cdot \nabla_{\mathbf{r}} \; \Psi_0^{(r > b_0)}(\mathbf{r}_0 + \mathbf{r} \mid \mathbf{q}_0)$$

on the surface of the spherical cavity where $r = b_0$. A solution of Poisson's equation which possesses the appropriate singularities and satisfies the boundary conditions is

(D 3)

$$\Psi_0^{(r > b_0)}(\mathbf{r}_0 + \mathbf{r} \mid \mathbf{q}_0) =$$

$$= \sum_{n=0}^{+\infty} \frac{2n + 1}{\left[(n + 1)\epsilon + n\,\epsilon_0\right]} \sum_{i=1}^{s_0} \frac{e_i\, r_i^n}{r^{(n+1)}} P_n(\cos \Theta_{r_i r})$$

$$\Psi_0^{(r < b_0)}(\mathbf{r}_0 + \mathbf{r} \mid \mathbf{q}_0) = \sum_{i=1}^{s_0} \frac{e_i}{\epsilon_0 \left|\mathbf{r} - \mathbf{r}_i\right|} +$$

$$+ \sum_{n=0}^{+\infty} \frac{(n + 1)(\epsilon_0 - \epsilon)}{\epsilon_0\left[(n + 1)\epsilon + n\,\epsilon_0\right]} \sum_{i=1}^{s_0} \frac{e_i\, r_i^n\, r^n}{b_0^{2n+1}} P_n(\cos \Theta_{r_i r})$$

where

$$\sum_{i=1}^{s_0} \frac{e_i}{\epsilon_0 |\mathbf{r} - \mathbf{r}_i|} = \sum_{n=0}^{+\infty} \sum_{i=1}^{s_0} \frac{e_i \, r_i^{n}}{\epsilon_0 \, r^{(n+1)}} P_n(\cos \Theta_{r_i r})$$

and

$$P_n(\cos \Theta_{r_i r}) = \tag{D 4}$$

$$= \sum_{m=-n}^{m=+n} \frac{(n - |m|)!}{(n + |m|)!} P_n^{|m|}(\cos \theta_i) \, P_n^{|m|}(\cos \theta) \, \exp\left[im(\varphi - \varphi_i) \right]$$

The potential at the point $\mathbf{r}_0 + \mathbf{r}$ due to an electrolyte ion of species α at $\mathbf{q}_\alpha$ and a dielectric sphere (uncharged non-electrolyte molecule, i.e., a non-electrolyte molecule with its electrostatic charging parameter ξ_0 equal to zero) centered at $\mathbf{r}_0$ will be denoted by $\Psi_{0\alpha}(\mathbf{r}_0 + \mathbf{r} \mid \mathbf{q}_0, \mathbf{q}_\alpha; \xi_0 = 0)$. It satisfies Poisson's equation

$$\nabla_{\mathbf{r}}^2 \; \Psi_{0\alpha}^{(r > b_0)}(\mathbf{r}_0 + \mathbf{r} \mid \mathbf{q}_0, \mathbf{q}_\alpha; \xi_0 = 0) = \tag{D 5}$$

$$= - \frac{4\pi e_\alpha}{\epsilon} \, \delta(\mathbf{r} - \mathbf{r}_{\alpha 0})$$

$$\nabla^2_{\underset{\sim}{\underline{r}}} \; \Psi_{o\alpha}^{(\underline{r} < \underline{b}_o)}(\underset{\sim}{\underline{r}}_o + \underset{\sim}{\underline{r}} \mid \underset{\sim}{\underline{q}}_o, \underset{\sim}{\underline{q}}_\alpha ; \xi_o = 0) = 0$$

and the boundary conditions

$$\Psi_{o\alpha}^{(\underline{r} < \underline{b}_o)}(\underset{\sim}{\underline{r}}_o + \underset{\sim}{\underline{r}} \mid \underset{\sim}{\underline{q}}_o, \underset{\sim}{\underline{q}}_\alpha ; \xi_o = 0) = \tag{D 6}$$

$$= \Psi_{o\alpha}^{(\underline{r} > \underline{b}_o)}(\underset{\sim}{\underline{r}}_o + \underset{\sim}{\underline{r}} \mid \underset{\sim}{\underline{q}}_o, \underset{\sim}{\underline{q}}_\alpha ; \xi_o = 0)$$

$$\underset{\sim}{\underline{1}}_{\underline{r}} \times \nabla_{\underset{\sim}{\underline{r}}} \; \Psi_{o\alpha}^{(\underline{r} < \underline{b}_o)}(\underset{\sim}{\underline{r}}_o + \underset{\sim}{\underline{r}} \mid \underset{\sim}{\underline{q}}_o, \underset{\sim}{\underline{q}}_\alpha ; \xi_o = 0) =$$

$$= \underset{\sim}{\underline{1}}_{\underline{r}} \times \nabla_{\underset{\sim}{\underline{r}}} \; \Psi_{o\alpha}^{(\underline{r} > \underline{b}_o)}(\underset{\sim}{\underline{r}}_o + \underset{\sim}{\underline{r}} \mid \underset{\sim}{\underline{q}}_o, \underset{\sim}{\underline{q}}_\alpha ; \xi_o = 0)$$

$$\epsilon_o \, \underset{\sim}{\underline{1}}_{\underline{r}} \cdot \nabla_{\underset{\sim}{\underline{r}}} \; \Psi_{o\alpha}^{(\underline{r} < \underline{b}_o)}(\underset{\sim}{\underline{r}}_o + \underset{\sim}{\underline{r}} \mid \underset{\sim}{\underline{q}}_o, \underset{\sim}{\underline{q}}_\alpha ; \xi_o = 0) =$$

$$= \epsilon \, \underset{\sim}{\underline{1}}_{\underline{r}} \cdot \nabla_{\underset{\sim}{\underline{r}}} \; \Psi_{o\alpha}^{(\underline{r} > \underline{b}_o)}(\underset{\sim}{\underline{r}}_o + \underset{\sim}{\underline{r}} \mid \underset{\sim}{\underline{q}}_o, \underset{\sim}{\underline{q}}_\alpha ; \xi_o = 0) .$$

on the surface of the spherical cavity where $\underline{r} = \underline{b}_o$. A solution of this equation is

(D 7)

$$\Psi_{0\alpha}^{(r > b_0)}(\underset{\sim}{r}_0 + \underset{\sim}{r} \mid \underset{\sim}{q}_0, \underset{\sim}{q}_\alpha; \xi_0 = 0) = \Psi_\alpha(\underset{\sim}{r}_0 + \underset{\sim}{r} \mid \underset{\sim}{q}_\alpha) +$$

$$+ \sum_{n=0}^{+\infty} \frac{n(\epsilon - \epsilon_0)\, e_\alpha}{\epsilon\left[(n+1)\epsilon + n\,\epsilon_0\right]} \frac{b_0^{2n+1}}{r^{n+1}\, r_{\alpha 0}^{n+1}} P_n(\cos \Theta_{r_{\alpha 0} r})$$

$$\Psi_{0\alpha}^{(r < b_0)}(\underset{\sim}{r}_0 + \underset{\sim}{r} \mid \underset{\sim}{q}_0, \underset{\sim}{q}_\alpha; \xi_0 = 0) =$$

$$= \sum_{n=0}^{+\infty} \frac{(2n+1)}{\left[(n+1)\epsilon + n\,\epsilon_0\right]} \frac{e_\alpha\, r^n}{r_{\alpha 0}^{n+1}} P_n(\cos \Theta_{r_{\alpha 0} r})$$

where

$$\Psi_\alpha(\underset{\sim}{r}_0 + \underset{\sim}{r} \mid \underset{\sim}{q}_\alpha) = \frac{e_\alpha}{\epsilon\left|(\underset{\sim}{r}_0 + \underset{\sim}{r}) - \underset{\sim}{r}_\alpha\right|} =$$

$$= \sum_{n=0}^{+\infty} \frac{e_\alpha}{\epsilon} \frac{r^n}{r_{\alpha 0}^{n+1}} P_n(\cos \Theta_{r_{\alpha 0} r})$$

is the potential at the point $\underset{\sim}{r}_0 + \underset{\sim}{r}$ due to an electrolyte ion of species α at $\underset{\sim}{q}_\alpha$.

From equations D 3 and D 7, the potential at the point $\underset{\sim}{\underline{r}}_o + \underset{\sim}{\underline{r}}$ due to a non-electrolyte molecule of species o at $\underset{\sim}{\underline{q}}_o$ and to an electrolyte ion of species α at $\underset{\sim}{\underline{q}}_\alpha$ is

(D 8)

$$\Psi_{o\alpha}^{(\underline{r} > \underline{b}_o)}(\underset{\sim}{\underline{r}}_o + \underset{\sim}{\underline{r}} \mid \underset{\sim}{\underline{q}}_o, \underset{\sim}{\underline{q}}_\alpha) = \Psi_o^{(\underline{r} > \underline{b}_o)}(\underset{\sim}{\underline{r}}_o + \underset{\sim}{\underline{r}} \mid \underset{\sim}{\underline{q}}_o) + $$

$$+ \Psi_{o\alpha}^{(\underline{r} > \underline{b}_o)}(\underset{\sim}{\underline{r}}_o + \underset{\sim}{\underline{r}} \mid \underset{\sim}{\underline{q}}_o, \underset{\sim}{\underline{q}}_\alpha; \xi_o = 0)$$

$$\Psi_{o\alpha}^{(\underline{r} < \underline{b}_o)}(\underset{\sim}{\underline{r}}_o + \underset{\sim}{\underline{r}} \mid \underset{\sim}{\underline{q}}_o, \underset{\sim}{\underline{q}}_\alpha) = \Psi_o^{(\underline{r} < \underline{b}_o)}(\underset{\sim}{\underline{r}}_o + \underset{\sim}{\underline{r}} \mid \underset{\sim}{\underline{q}}_o) + $$

$$+ \Psi_{o\alpha}^{(\underline{r} < \underline{b}_o)}(\underset{\sim}{\underline{r}}_o + \underset{\sim}{\underline{r}} \mid \underset{\sim}{\underline{q}}_o, \underset{\sim}{\underline{q}}_\alpha; \xi_o = 0) .$$

The electrostatic contribution to $\underline{V}_{o\alpha}(\underset{\sim}{\underline{q}}_o, \underset{\sim}{\underline{q}}_\alpha)$ is equal to the electrostatic work of charging the ion of species α with the system in the given configuration minus the electrostatic work of charging the ion of species α when it is an infinite distance from the non-electrolyte molecule of species o . Upon introducing ξ_α, the electrostatic charging parameter of the ion of species α at $\underset{\sim}{\underline{q}}_\alpha$, we can express this as

(D 9)

$$\int_0^1 \Psi_{o\alpha}^{(\underline{r}_{\alpha o} > \underline{b}_o)}(\underset{\sim}{\underline{r}}_\alpha \mid \underset{\sim}{\underline{q}}_o, \underset{\sim}{\underline{q}}_\alpha; \xi_\alpha) \ \underline{e}_\alpha \, \underline{d}\xi_\alpha$$

$$-\int_0^1 \Psi_\alpha(\underset{\sim}{r}_\alpha \mid \underset{\sim}{q}_\alpha;\ \xi_\alpha)\ e_\alpha\ d\xi_\alpha\ .$$

If $r_{\alpha o} \geq a_{\alpha o}$, the electrostatic contribution to $V_{o\alpha}(\underset{\sim}{q}_o, \underset{\sim}{q}_\alpha)$ as obtained by carrying out the above integrations is

$$V_{o\alpha}(\underset{\sim}{q}_o, \underset{\sim}{q}_\alpha) = V_{o\alpha}^{LR}(\underset{\sim}{q}_o, \underset{\sim}{q}_\alpha) + V_{o\alpha}^{SR}(\underset{\sim}{q}_o, \underset{\sim}{q}_\alpha) \qquad \text{(D 10)}$$

where

$$V_{o\alpha}^{LR}(\underset{\sim}{q}_o, \underset{\sim}{q}_\alpha) = e_\alpha\ \Psi_o^{(r_{\alpha o} > b_o)}(\underset{\sim}{r}_\alpha \mid \underset{\sim}{q}_o) \qquad (r_{\alpha o} \geq a_{\alpha o})\ . \qquad \text{(D 11)}$$

$$= \sum_{i=1}^{s_o} e_i\ \Psi_{o\alpha}^{(r_{\alpha o} > b_o)}(\underset{\sim}{r}_o + \underset{\sim}{r}_i \mid \underset{\sim}{q}_o, \underset{\sim}{q}_\alpha;\ \xi_o = 0)$$

is the potential of the long range Coulombic force acting between the non-electrolyte molecule electrolyte ion pair; and where

$$V_{o\alpha}^{SR}(\underset{\sim}{q}_o, \underset{\sim}{q}_\alpha) = -\sum_{n=4}^{+\infty} d_{o\alpha,n} \left(\frac{a_{\alpha o}}{r_{\alpha o}}\right)^n \qquad (r_{\alpha o} \geq a_{\alpha o}) \qquad \text{(D 12)}$$

with

(D 13)

$$\underline{d}_{o\alpha,\underline{n}} = -\underline{e}_\alpha^2 \frac{(\underline{n}-2)(\epsilon-\epsilon_o)}{2\epsilon\left[\underline{n}\epsilon + (\underline{n}-2)\epsilon_o\right]} \frac{\underline{b}_o^{\underline{n}-1}}{\underline{a}_{\alpha o}^{\underline{n}}} \qquad (\underline{n} \text{ even})$$

$$\underline{d}_{o\alpha,\underline{n}} = 0 \qquad (\underline{n} \text{ odd})$$

represents a short range ion-cavity repulsive force between the electrolyte ion and an image distribution in the spherical cavity of low dielectric constant created by the non-electrolyte molecule in the solvent.

When the non-electrolyte molecule possesses a charge distribution which may be characterized by a point or ideal dipole of moment $\underset{\sim}{\mu}_o = \mu_o \underset{\approx}{k}_o = \mu_o \underset{\approx}{1}_{\underline{R}_o}$ located at its center, the potential as given by equation D 3 degenerates into

(D 14)

$$\Psi_o^{(\underline{r} > \underline{b}_o)}(\underset{\approx}{r}_o + \underset{\approx}{r} \mid \underset{\approx}{q}_o) = \frac{3}{2\epsilon + \epsilon_o} \frac{1}{\underline{r}^2} \underset{\sim}{\mu}_o \cdot \underset{\approx}{1}_{\underline{r}} .$$

Then equation D 11 becomes

(D 15)

$$\underline{V}_{o\alpha}^{LR}(\underset{\sim}{q}_o, \underset{\sim}{q}_\alpha) = \frac{3\underline{e}_\alpha}{2\epsilon + \epsilon_o} \frac{1}{\underline{r}_{\alpha o}^2} \underset{\sim}{\mu}_o \cdot \underset{\approx}{1}_{\underline{r}_{\alpha o}} \qquad (\underline{r}_{\alpha o} \geq \underline{a}_{\alpha o})$$

From equation D 10 the force on a dipolar ion of species o at $\underline{q}_o$ due to an electrolyte ion of species α at $\underline{q}_\alpha$ is

(D 16)

$$\underline{F}_{\alpha o}(\underline{q}_\alpha, \underline{q}_o) = - \underline{F}_{o\alpha}(\underline{q}_o, \underline{q}_\alpha) = - \nabla_{\underline{r}_o} \underline{V}_{o\alpha}(\underline{q}_o, \underline{q}_\alpha) =$$

$$= \left[\sum_{\underline{n}=4}^{+\infty} \underline{n}\, \underline{d}_{o\alpha,\underline{n}} \frac{\underline{a}_{\alpha o}^{\underline{n}}}{\underline{r}_{\alpha o}^{\underline{n}+1}} - \frac{9\underline{e}_\alpha}{2\epsilon + \epsilon_o} \frac{1}{\underline{r}_{\alpha o}^3} \underline{\mu}_o \cdot \underline{1r}_{\alpha o} \right] \underline{1r}_{\alpha o}$$

$$+ \left[\frac{3\underline{e}_\alpha}{2\epsilon + \epsilon_o} \frac{1}{\underline{r}_{\alpha o}^3} \right] \underline{\mu}_o .$$

The torque on the dipolar ion of species o at $\underline{q}_o$ due to an electrolyte ion of species α at $\underline{q}_\alpha$, as obtained from equation D 10, is

(D 17)

$$\underline{T}_{\alpha o}(\underline{q}_\alpha, \underline{q}_o) = - \nabla_{\underline{R}_o} \underline{V}_{o\alpha}(\underline{q}_o, \underline{q}_\alpha) =$$

$$= - \frac{3\underline{e}_\alpha}{2\epsilon + \epsilon_o} \frac{1}{\underline{r}_{\alpha o}^2} \underline{\mu}_o \times \underline{1r}_{\alpha o} .$$

The Potential Energy of a Non-electrolyte Molecule in a Homogeneous Electric Field

First we shall consider the case of a dielectric sphere (an uncharged non-electrolyte molecule) under the influence of outside sources leading to a constant, uniform, parallel external field $\underset{\sim}{E} = E\,\underset{\sim}{k}$ at large distances from its center at $\underset{\sim}{r}_0$. The potential at any point $\underset{\sim}{r}_0 + \underset{\sim}{r}$ is given by[47]

(D 18)

$$\psi_0^{(\underset{\sim}{E};\, r < b_0)}(\underset{\sim}{r}_0 + \underset{\sim}{r} \mid \underset{\sim}{q}_0;\, \xi_0 = 0) = -\frac{3\epsilon}{2\epsilon + \epsilon_0}\, r\, \underset{\sim}{E} \cdot \underset{\sim}{1}_r$$

$$\psi_0^{(\underset{\sim}{E};\, r > b_0)}(\underset{\sim}{r}_0 + \underset{\sim}{r} \mid \underset{\sim}{q}_0;\, \xi_0 = 0) = -r\, \underset{\sim}{E} \cdot \underset{\sim}{1}_r + \frac{\epsilon_0 - \epsilon}{2\epsilon + \epsilon_0}\, \frac{b_0^3}{r^2}\, \underset{\sim}{E} \cdot \underset{\sim}{1}_r$$

and the corresponding electric field intensities are

(D 19)

$$\underset{\sim}{E}_0^{(r < b_0)}(\underset{\sim}{r}_0 + \underset{\sim}{r} \mid \underset{\sim}{q}_0;\, \xi_0 = 0) = \frac{3\epsilon}{2\epsilon + \epsilon_0}\, \underset{\sim}{E}$$

$$\underset{\sim}{E}_0^{(r > b_0)}(\underset{\sim}{r}_0 + \underset{\sim}{r} \mid \underset{\sim}{q}_0;\, \xi_0 = 0) = \underset{\sim}{E} + \frac{\epsilon - \epsilon_0}{2\epsilon + \epsilon_0}\, \frac{b_0^3}{r^3} \left[\underset{\sim}{E} - 3\, \underset{\sim}{E} \cdot \underset{\sim}{1}_r\, \underset{\sim}{1}_r\right].$$

Consequently, the potential energy of a non-electrolyte molecule of species o at $\underline{\underset{\sim}{q}}_o$ in the homogeneous electric field $\underline{\underset{\sim}{E}}$ is given by

(D 20)

$$\underline{V}_o^{(\underline{E})}(\underline{\underset{\sim}{q}}_o) = \sum_{\underline{i}=1}^{\underline{s}_o} \underline{e}_{\underline{i}} \; \Psi_o^{(\underline{E};\underline{r}_{\underline{i}} < \underline{b}_o)}(\underline{\underset{\sim}{r}}_o + \underline{\underset{\sim}{r}}_{\underline{i}} \mid \underline{\underset{\sim}{q}}_o ; \xi_o = 0) =$$

$$= - \frac{3\epsilon}{2\epsilon + \epsilon_o} \underline{\underset{\sim}{E}} \cdot \left(\sum_{\underline{i}=1}^{\underline{s}_o} \underline{e}_{\underline{i}} \underline{\underset{\sim}{r}}_{\underline{i}} \right).$$

The force on a non-electrolyte molecule of species o at $\underline{\underset{\sim}{q}}_o$ due to the external field $\underline{\underset{\sim}{E}}$ is

(D 21)

$$\underline{\underset{\sim}{F}}_o^{(\underline{E})}(\underline{\underset{\sim}{q}}_o) = - \nabla_{\underline{\underset{\sim}{r}}_o} \underline{V}_o^{(\underline{E})}(\underline{\underset{\sim}{q}}_o) = \sum_{\underline{i}=1}^{\underline{s}_o} \underline{e}_{\underline{i}} \underline{\underset{\sim}{E}}_o^{(\underline{r}_{\underline{i}} < \underline{b}_o)}(\underline{\underset{\sim}{r}}_o + \underline{\underset{\sim}{r}}_{\underline{i}} \mid \underline{\underset{\sim}{q}}_o ; \xi_o = 0)$$

$$= \frac{3\epsilon}{2\epsilon + \epsilon_o} \left(\sum_{\underline{i}=1}^{\underline{s}_o} \underline{e}_{\underline{i}} \right) \underline{\underset{\sim}{E}} = 0.$$

The torque on a non-electrolyte molecule of species o at

$\underset{\sim}{q}_o$ due to the external field $\underset{\sim}{E}$ is

$$\underset{\sim}{T}_o^{(E)}(\underset{\sim}{q}_o) = - \nabla_{\underset{\sim}{R}_o} V_o^{(E)}(\underset{\sim}{q}_o) = \tag{D 22}$$

$$= \sum_{i=1}^{s_o} \underset{\sim}{r}_i \times e_i \underset{\sim}{E}_o^{(r_i < b_o)}(\underset{\sim}{r}_o + \underset{\sim}{r}_i \mid \underset{\sim}{q}_o ; \xi_o = 0) =$$

$$= \frac{3\epsilon}{2\epsilon + \epsilon_o} \left(\sum_{i=1}^{s_o} e_i \underset{\sim}{r}_i \right) \times \underset{\sim}{E} \; .$$

The potential energy of a dipolar non-electrolyte molecule of species o at $\underset{\sim}{q}_o$ in the homogeneous electric field $\underset{\sim}{E}$ is given by

$$V_o^{(E)}(\underset{\sim}{q}_o) = - \frac{3\epsilon}{2\epsilon + \epsilon_o} \underset{\sim}{E} \cdot \underset{\sim}{\mu}_o \; . \tag{D 23}$$

The force on the dipolar ion due to the external field $\underset{\sim}{E}$ is

$$\underset{\sim}{F}_o^{(E)}(\underset{\sim}{q}_o) = 0 \; . \tag{D 24}$$

Finally, the torque on the dipolar ion due to the external

field $\underline{E}$ is

$$\underline{T}_o^{(\underline{E})}(\underline{q}_o) = \frac{3\epsilon}{2\epsilon + \epsilon_o} \underline{\mu}_o \times \underline{E} . \qquad \text{(D 25)}$$

The Potential Energy of an Electrolyte Ion in a Homogeneous Electric Field

The potential energy of an electrolyte ion of species α at $\underline{q}_\alpha$ in the homogeneous electric field $\underline{E}$ is given by

$$\underline{V}_\alpha^{(\underline{E})}(\underline{q}_\alpha) = - \underline{e}_\alpha \, \underline{r}_{\alpha o} \, \underline{E} \cdot \underline{1}_{\underline{r}_{\alpha o}} \qquad \text{(D 26)}$$

and the force on the electrolyte ion due to the external field $\underline{E}$ is

$$\underline{F}_\alpha^{(\underline{E})}(\underline{q}_\alpha) = - \nabla_{\underline{r}_{\alpha o}} \underline{V}_\alpha^{(\underline{E})}(\underline{q}_\alpha) = \underline{e}_\alpha \, \underline{E} . \qquad \text{(D 27)}$$

If a non-electrolyte molecule of species o is at $\underline{q}_o$, this potential energy becomes

(D 28)

$$\underline{V}_{\alpha|o}^{(\underline{E})}(\underline{q}_\alpha | \underline{q}_o ; \xi_o = 0) = \underline{e}_\alpha \, \Psi_o^{(\underline{E};\ \underline{r}_{\alpha o} > \underline{b}_o)}(\underline{r}_\alpha | \underline{q}_o ; \xi_o = 0) =$$

$$= - e_\alpha r_{\alpha o} \mathbf{E} \cdot \mathbf{1}_{r_{\alpha o}} + e_\alpha \frac{\epsilon_o - \epsilon}{2\epsilon + \epsilon_o} \frac{b_o^3}{r_{\alpha o}^2} \mathbf{E} \cdot \mathbf{1}_{r_{\alpha o}}$$

and the force on the electrolyte ion of species α due to the external field becomes

(D 29)

$$\mathbf{F}_{\alpha|o}^{(E)}(\mathbf{q}_\alpha|\mathbf{q}_o;\ \xi_o = 0) = - \nabla_{\mathbf{r}_{\alpha o}} V_{\alpha|o}^{(E)}(\mathbf{q}_\alpha|\mathbf{q}_o;\ \xi_o = 0) =$$

$$= e_\alpha \mathbf{E} + e_\alpha \frac{\epsilon - \epsilon_o}{2\epsilon + \epsilon_o} \frac{b_o^3}{r_{\alpha o}^3} \left(\mathbf{E} - 3\, \mathbf{E} \cdot \mathbf{1}_{r_{\alpha o}} \mathbf{1}_{r_{\alpha o}} \right) .$$

The last term arises as a result of having a spherical cavity of low dielectric constant created by the non-electrolyte molecule in the solvent. It may be thought of as due to a dipole of moment $\boldsymbol{\mu}_o^{(ind)} = \left[(\epsilon_o - \epsilon)/3\right] b_o^3 \mathbf{E}$ induced in the non-electrolyte molecule and hence included in $\mathbf{F}_{o\alpha}(\mathbf{q}_o, \mathbf{q}_\alpha)$ but not in $\mathbf{F}_{\alpha o}(\mathbf{q}_o, \mathbf{q}_\alpha)$. For this contribution to $\mathbf{F}_{o\alpha}(\mathbf{q}_o, \mathbf{q}_\alpha)$ we use the notation

(D 30)

$$\mathbf{F}_{o\alpha}^{(E)}(\mathbf{q}_o, \mathbf{q}_\alpha;\ \xi_o = 0) = - \nabla_{\mathbf{r}_{\alpha o}} V_{o\alpha}^{(E)}(\mathbf{q}_o, \mathbf{q}_\alpha;\ \xi_o = 0) =$$

$$= e_\alpha \frac{\epsilon - \epsilon_o}{2\epsilon + \epsilon_o} \frac{b_o^3}{r_{\alpha o}^3} \left(\mathbf{E} - 3\, \mathbf{E} \cdot \mathbf{1}_{r_{\alpha o}} \mathbf{1}_{r_{\alpha o}} \right) .$$

Appendix E: Values of the Function $G(x)$

Table VI

Values of the Function $G(x)$

$$G(x) = \frac{13 - 3x - \frac{13}{6}x^2 - \frac{7}{6}x^3 + \frac{1}{2}x^4 e^x Ei(x)}{6 + 6x + \frac{8}{3}x^2 + \frac{2}{3}x^3}$$

x	$G(x)$	x	$G(x)$
0.00	2.167	1.80	- 0.1422
0.10	1.913	1.90	- 0.1923
0.20	1.683	2.00	- 0.2385
0.30	1.474	2.10	- 0.2814
0.40	1.284	2.20	- 0.3208
0.50	1.106	2.30	- 0.3575
0.60	0.955	2.40	- 0.3915
0.70	0.812	2.50	- 0.4231
0.80	0.682	2.60	- 0.4525
0.90	0.563	2.70	- 0.4798
1.00	0.454	2.80	- 0.5053
1.10	0.3551	2.90	- 0.5291
1.20	0.2644	3.00	- 0.5513
1.30	0.1813	3.10	- 0.5721
1.40	0.1052	3.20	- 0.5916
1.50	0.0354	3.30	- 0.6098
1.60	- 0.0288	3.40	- 0.6269
1.70	- 0.0878	3.50	- 0.6431

Appendix F: Values of the Functions $H(x)$, $K(x)$, and $L(x)$

Table VII

Values of the Function $H(x)$

$$H(x) = -\frac{2}{x^2} - \frac{2}{x^4} \ln (1 + x)(1 - x)$$

x	$H(x)$	x	$H(x)$
0.00	1.000	0.65	1.418
0.10	1.007	0.70	1.527
0.20	1.028	0.75	1.670
0.30	1.064	0.80	1.864
0.40	1.121	0.85	2.143
0.50	1.206	0.90	2.593
0.55	1.262	0.95	3.500
0.60	1.332		

Table VIII

Values of the Function $K(x)$

$$K(x) = \frac{14}{3x^4}\left[\left(1 + \frac{5}{2x^3}\right)\ln(1 + x) + \left(1 - \frac{5}{2x^3}\right)\ln(1 - x) - \frac{5}{x^2} - \frac{5}{3}\right]$$

x	$K(x)$	x	$K(x)$
0.00	1.000	0.65	1.714
0.10	1.015	0.70	1.920
0.20	1.043	0.75	2.199
0.30	1.102	0.80	2.596
0.40	1.194	0.85	3.202
0.50	1.336	0.90	4.248
0.55	1.434	0.95	6.576
0.60	1.557		

Table IX

Values of the Function $L(x)$

$$L(x) = \frac{2}{1 - x^2} + \frac{2}{x^2} + \frac{2}{x^4}\ \ln\,(1 + x)(1 - x)$$

x	$L(x)$	x	$L(x)$
0.00	1.000	0.65	2.045
0.10	1.013	0.70	2.394
0.20	1.056	0.75	2.902
0.30	1.133	0.80	3.692
0.40	1.260	0.85	5.064
0.50	1.461	0.90	7.933
0.55	1.605	0.95	17.013
0.60	1.793		

Appendix G: Important Physical Constants

Table X

Important Physical Constants[a]

Faraday	96,493.1 absolute coulomb equivalent^{-1}
	2.89277×10^{14} absolute electrostatic units equivalent^{-1}
Avogadro number	6.02380×10^{23}
Protonic charge (*e*)	4.80223×10^{-10} electrostatic unit
	1.601864×10^{-19} absolute coulomb
Boltzmann's constant (*k*)	1.380257×10^{-16} erg degree^{-1} molecule^{-1}
0° C.	273.160° absolute

[a]R. A. Robinson and R. H. Stokes, "Electrolyte Solutions," 2nd ed, Butterworth and Co., London, 1965, p xv.

Appendix H: Physical Properties of Water

Table XI

Physical Properties of Water[a]

	$\underline{T}$ = 0° C	$\underline{T}$ = 25° C
Dielectric constant (ϵ)	87.740	78.303
Density in grams milliliter^{-1}	0.99987	0.99707
Viscosity (η) in poise	17.87×10^{-3}	8.903×10^{-3}

[a]R. A. Robinson and R. H. Stokes, "Electrolyte Solutions," 2nd ed, Butterworth and Co., London, 1965, p 457.

Appendix I: Physical Properties of Electrolyte Ions

The limiting equivalent ionic conductances (λ cm^2 ohm^{-1} equivalent^{-1}) of selected ions in water at infinite dilution are given in Table XII. The limiting ionic translational diffusion coefficients ($\underline{D}$ cm^2/sec) were computed from these conductances with the equation

$$\underline{D} = \frac{\underline{k}\,\underline{T}\,\underline{N}\,\lambda}{|\underline{Z}|\,\underline{F}^2} = 8.929 \times 10^{-10}\,\frac{\lambda\,\underline{T}}{|\underline{Z}|}\ \text{cm}^2\ \text{sec}^{-1} \qquad \text{(I 1)}$$

where $\underline{Z}$ is the valency of the ion; where $\underline{N}$ is the Avogadro number; and $\underline{F}$ is the Faraday. These various physical constants are given in Appendix G.

Table XIII contains crystallographic radii of the ions; Stokes' law radii, calculated with equation 73; and for the tetraalkylammonium ions, radii estimated from molecular volumes or models.

Table XII

Limiting Equivalent Conductances and Translational Diffusion Coefficients of Ions in Water[a]

	$\underline{T}$ = 0° C		$\underline{T}$ = 25° C	
Ion	λ	$\underline{D} \times 10^6$	λ	$\underline{D} \times 10^6$
H^+	225	54.9	349.81	93.13
OH^-	105	25.6	198.3	52.79
Li^+	19.4	4.73	38.68	10.298
Na^+	26.5	6.46	50.10	13.338
K^+	40.7	9.93	73.50	19.568
Cs^+	44	10.7	77.26	20.569
NH_4^+	40.2	9.80	73.55	19.581
$N(CH_3)_4^+$	24.1	5.88	44.92	11.959
$N(C_2H_5)_4^+$	16.4	4.00	32.66	8.695
$N(C_3H_7)_4^+$	11.5	2.80	23.42	6.235
$N(C_4H_9)_4^+$	9.6	2.34	19.47	5.183
F^-	-	(7.26)	55.4	14.75
Cl^-	41.0	10.00	76.35	20.327
I^-	41.4	10.10	76.84	20.457
Ba^{++}	34.0	4.15	63.63	8.470

[a] R. A. Robinson and R. H. Stokes, "Electrolyte Solutions," 2nd ed, Butterworth and Co., London, 1965, p 465.

Table XIII

Radii of Ions (in Angström Units)

Ion	b (crystallographic)[a]	b (model)[b]	ℛ (Stokes, 0° C)	ℛ (Stokes, 25° C)
H^+	-	-	0.204	0.263
OH^-	-	-	0.437	0.465
Li^+	0.607	-	2.37	2.38
Na^+	0.958	-	1.73	1.84
K^+	1.331	-	1.13	1.25
Cs^+	1.656	-	1.05	1.19
NH_4^+	1.48	-	1.14	1.25
$N(CH_3)_4^+$	-	3.47	1.90	2.05
$N(C_2H_5)_4^+$	-	4.00	2.80	2.82
$N(C_3H_7)_4^+$	-	4.52	4.00	3.93
$N(C_4H_9)_4^+$	-	4.94	4.79	4.73
F^-	1.341	-	-	1.66
Cl^-	1.806	-	1.12	1.21
I^-	2.168	-	1.11	1.20
Ba^{++}	1.35	-	2.70	2.90

[a] H. S. Harned and B. B. Owen, "Physical Chemistry of Electrolytic Solutions," Reinhold Publishing Corporation, New York, 1958, p. 164.

[b] R. A. Robinson and R. H. Stokes, "Electrolyte Solutions," 2nd ed, Butterworth and Co., London, 1965, p. 125.

Appendix J: Physical Properties of Non-electrolyte Molecules

Values of the dipole moments (Table XIV), the limiting translational diffusion coefficients (Table XV), the limiting rotational diffusion coefficients (Table XVI), and the radii (Table XVII) of selected non-electrolyte molecules are listed in this appendix.

For the molecular radii, estimates obtained by several different methods are given. We have tabulated: (1) radii calculated from the van der Waals volumes of the molecules obtained by summing atomic increments estimated from crystallographic data[48]; (2) radii estimated from molecular models; (3) hydrodynamic radii calculated from the limiting translational diffusion coefficients given in Table XV using Stokes' Law, equation 73; (4) hydrodynamic radii calculated from the limiting (apparent or partial) molal volumes given in Table XVIII; and (5) hydrodynamic radii calculated from the limiting molar intrinsic viscosities given in Table XIX using Einstein's equation relating the viscosity of a solution to the volume fraction of the solute.[49] Those radii used in calculating the mass transport ratios given in Tables IV and V are indicated by the asterisk (*).

Numbers enclosed in parentheses ( ) are estimates based on the best available experimental data.

Table XIV

Dipole Moments of Non-electrolyte

Molecules (in Debye Units)

Molecule	Physical State	Dipole Moment(μ_0)
Acetone	Gas	2.85[a]
Ethyl Acetate	Gas	1.76[b]
Glycine	Solution	15.5[c]
Maltose	-	(3)
d-Mannitol	Solution	3.9[d]
Raffinose	-	(3)
Resorcinol	Solution	2.07[e]
Sucrose	Solution	3[f]
Urea	Solution	4.5[g]

[a]C. T. Zahn, Physik Z., **33**, 686 (1932).

[b]C. T. Zahn, Physik Z., **33**, 730 (1932).

[c]J. T. Edsall and J. Wyman, "Biophysical Chemistry," Academic Press Inc., New York, 1958, pp 371-373.

[d]P. Girard and P. Abadie, Compt. rend., **197**, 146 (1933).

[e]J. J. Lander and W. J. Svirbely, J. Amer. Chem. Soc., **67**, 322 (1945).

[f]E. Landt, Naturwissenschaften, **22**, 809 (1934).

[g]W. R. Gilkerson and K. K. Srivastava, J. Phys. Chem., **64**, 1485 (1960).

Table XV

Limiting Translational Diffusion Coefficients of Non-electrolyte Molecules in Water at Infinite Dilution

Molecule	$D_o \times 10^6$ at 0° C	$D_o \times 10^6$ at 25° C
Acetone	-	13.0[a]
Ethyl Acetate	-	(13)[b]
Glycine	-	10.635[c]
Maltose	-	(5.0)[d]
d-Mannitol	(2.62)[e]	-
Raffinose	(1.93)[f]	4.359[g]
Resorcinol	(4.0)[d]	-
Sucrose	(2.33)[h]	-
Urea	(6.62)[i]	13.817[j]

[a]D. K. Anderson, J. R. Hall, and A. L. Babb, J. Phys. Chem., **62**, 404 (1958).

[b]J. B. Lewis, J. Appl. Chem. (London), **5**, 228 (1955).

[c]M. S. Lyons and J. V. Thomas, J. Amer. Chem. Soc., **72**, 4506 (1950).

[d]A. E. Stearn, E. M. Irish, and H. Eyring, J. Phys. Chem., **44**, 981 (1940).

[e]K. G. Stern, S. Singer, and S. Davis, J. Biol. Chem., **167**,321 (1947).

[f]L. G. Longsworth, J. Amer. Chem. Soc., **74**, 4155 (1952).

[g]P. J. Dunlop, J. Phys. Chem., **60**, 1464 (1956).

[h]L. J. Gosting and M. S. Morris, J. Amer. Chem. Soc., **71**, 1998 (1949).

[i]L. G. Longsworth, J. Phys. Chem., **58**, 770 (1954).

[j]L. J. Gosting and D. F. Akeley, J. Amer. Chem. Soc., **74**, 2058 (1952).

Table XVI

Limiting Rotational Diffusion Coefficients of Non-electrolyte Molecules in Water at Infinite Dilution

Molecule	$D_o^{(rot)} \times 10^{-10}$ at 0° C	$D_o^{(rot)} \times 10^{-10}$ at 25° C
Acetone	-	(0.695)[a]
Ethyl Acetate	-	(0.885)[a]
Glycine	-	1.02[b]
Maltose	-	(0.156)[a]
d-Mannitol	(0.108)[a]	-
Raffinose	(0.043)[a]	(0.103)[a]
Resorcinol	(0.355)[a]	-
Sucrose	(0.076)[a]	-
Urea	(0.478)[a]	(1.050)[a]

[a]An estimate computed from the hydrodynamic radii given in Table XVII using Stokes' Law, equation 81 .

[b]W. P. Conner and C. P. Smyth, J. Amer. Chem. Soc., **64**, 1870 (1942).

Table XVII

Radii of Non-electrolyte Molecules (in Angström Units)

Molecule	b_o		$\mathcal{R}_o$ at 0° C			$\mathcal{R}_o$ at 25° C		
	At. Inc.	Model	From D_o	From v_o	From $[\eta/o]$	From D_o	From v_o	From $[\eta/o]$
Acetone	2.48	-	-	-	-	1.89	2.98*	-
Ethyl Acetate	2.75*	-	-	-	-	1.9	-	-
Glycine	2.40	2.8[a]	-	-	2.53	2.31	2.58	2.82*
Maltose	4.00	4.5*	-	-	-	4.9	-	-
d-Mannitol	3.32	4.0[b]*	4.27	-	-	-	-	4.3[b]
Raffinose	4.56	-	5.79	-	-	5.63	4.95*	6.02
Resorcinol	2.87*	-	2.8	-	-	-	-	-
Sucrose	4.00	4.5[b]*	4.80	4.35	5.32	-	4.37	5.14
Urea	2.36	2.3[b]	1.69	-	-	1.77	2.60*	1.80

[a] J. T. Edsall and J. Wyman, "Biophysical Chemistry," Academic Press Inc., New York, 1958, p 296.

[b] S. G. Schultz and A. K. Solomon, J. Gen. Physiol., **44**, 1189 (1961).

Table XVIII

Limiting Molal Volumes of Non-electrolytes in Water at Infinite Dilution

Molecule	$\underline{v}_0$ at 0° C	$\underline{v}_0$ at 25° C
Acetone	-	67.04[a]
Glycine	-	43.20[b]
Raffinose	-	307.24[c]
Sucrose	(207.8)[d]	211.5[d]
Urea		44.22[e]

[a]C. Drucker, Arkiv. Kemi, Mineral. Geol., **14 A**, No. 15 (1941).

[b]F. T. Gucker, Jr., W. L. Ford, and C. E. Moser, J. Phys. Chem., **43**, 153 (1939).

[c]P. J. Dunlop, J. Phys. Chem., **60**, 1464 (1956).

[d]L. J. Gosting and M. S. Morris, J. Amer. Chem. Soc., **71**, 1998 (1949).

[e]F. T. Gucker, Jr., F. W. Gage, and C. E. Moser, J. Amer. Chem. Soc., **60**, 2582 (1938).

Table XIX

Limiting Molar Intrinsic Viscosities of Non-electrolytes in Water at Infinite Dilution

Molecule	$[\eta_o]$ at 0° C	$[\eta_o]$ at 25° C
Glycine	(0.1024)[a]	0.14235[b]
Raffinose	-	1.3831[c]
Sucrose	(0.9526)[d]	0.8558[d]
Urea	-	0.0370[e]

[a]M. S. Lyons and J. V. Thomas, J. Amer. Chem. Soc., **72**, 4506 (1950).

[b]L. S. Mason, P. M. Kampmeyer, and A. L. Robinson, J. Amer. Chem. Soc., **74**, 1287 (1952).

[c]P. J. Dunlop, J. Phys. Chem., **60**, 1464 (1956).

[d]L. J. Gosting and M. S. Morris, J. Amer. Chem. Soc., **71**, 1998 (1949).

[e]L. J. Gosting and D. F. Akeley, J. Amer. Chem. Soc., **74**, 2058 (1952).

Footnotes

(1) This paper is based upon part of a dissertation (directed by the late Professor John G. Kirkwood and by Professor Lars Onsager) submitted by Ray F. Snipes to the Faculty of the Graduate School of Yale University in partial fulfillment of the requirements for the Ph.D. degree, awarded June 1963.

(2) National Science Foundation Predoctoral Fellow, 1956-1959.

(3) D. A. MacInnes, "The Principles of Electrochemistry," Dover Publications,Inc., New York, 1961, pp. 91-95.

(4) L. G. Longsworth, J. Amer. Chem. Soc., **69**, 1288 (1947).

(5) J. H. Irving and J. G. Kirkwood, J. Chem. Phys., **18**, 817 (1950).

(6) R. J. Bearman and J. G. Kirkwood, J. Chem. Phys., **28**, 136 (1958).

(7) M. Y. Bearman and R. J. Bearman, J. Chem. Phys., **52**, 3189 (1970).

(8) J. S. Dahler, J. Chem. Phys., **30**, 1447 (1959).

(9) D. W. Condiff and J. S. Dahler, J. Chem. Phys., **44**, 3988 (1966).

(10) R. M. Fuoss and F. Accascina, "Electrolytic Conductance," Interscience Publishers,Inc., New York, 1959.

(11) J. G. Kirkwood, J. Chem. Phys., **2**, 351 (1934).

(12) F. A. Long and W. F. McDevit, Chem. Revs., **51**, 119 (1952).

(13) D. D. Fitts, "Non-equilibrium Thermodynamics," McGraw-Hill, Inc., New York, 1962, p 75.

(14) See Reference 10.

(15) For an extension of this theory to higher electrolyte concentrations (while neglecting short range forces), see: R. F. Snipes, Ph. D. Dissertation, Yale University, New Haven, Conn., 1963.

(16) H. Margenau and G. M. Murphy, "The Mathematics of Physics and Chemistry," D. Van Nostrand, Inc., Princeton, New Jersey, 1956, p 177.

(17) H. Goldstein, "Classical Mechanics," Addison-Wesley, Inc., Cambridge, Mass., 1950, pp 107-109.

(18) J. Willard Gibbs, "Elementary Principles in Statistical Mechanics," Yale University Press, New Haven, Conn., 1902, p 59.

(19) Given vectors $\mathbf{a}$ and $\mathbf{b}$ in real Euclidean 3-space R^3, we define a "quotient" $\mathbf{a} / \mathbf{b}$ only if $\mathbf{b} \neq \mathbf{0}$ and $\mathbf{a} = c\,\mathbf{b}$ where c is a real number. In this case, we define $\mathbf{a}/\mathbf{b} = c$. Thus if $c > 0$, the vectors $\mathbf{a}$ and $\mathbf{b}$ have the "same directions"; and if $c < 0$, they have "opposite directions."

(20) C. F. Curtiss, J. Chem. Phys., **24**, 225 (1956).

(21) T. L. Hill, "Statistical Mechanics," McGraw-Hill, Inc., New York, 1956.

(22) W. A. Steele, J. Chem. Phys., 39, 3197 (1963).

(23) J. G. Kirkwood, J. Chem. Phys., 3, 300 (1935).

(24) See Reference 21, pp 198-204.

(25) See Reference 20.

(26) R. M. Fuoss and L. Onsager, J. Phys. Chem., 62, 1339 (1958).

(27) A. Einstein, "Investigations on the Theory of the Brownian Movement," Dover Publications, Inc., New York, 1956, p 58.

(28) See Reference 6.

(29) See Reference 26.

(30) See Reference 11.

(31) J. G. Kirkwood, Chem. Revs., 24, 233 (1939).

(32) E. Jahnke and F. Emde, "Tables of Functions," Dover Publications, Inc., New York, 1945, pp 1-9. Our $Ei(x)$ is the $-Ei(-x)$ of these tables; and our $-Ei(-x)$ is the $\overline{Ei}(x)$ of these tables.

(33) See Reference 10, p 165.

(34) A. M. Squires, Ph. D. Thesis, Cornell University, Ithaca, N.Y., 1947.

(35) See Reference 3.

(36) E. W. Washburn, J. Amer. Chem. Soc., 31, 322 (1909).

(37) E. W. Washburn and E. B. Millard, J. Amer. Chem. Soc., 37, 694 (1915).

(38) M. Taylor and E. W. Sawyer, J. Chem. Soc., 2095 (1929).

(39) G. Davies, N. J. Hassid, and M. Taylor, J. Chem. Soc., 2497 (1932).

(40) C. H. Hale and T. De Vries, J. Amer. Chem. Soc., **70**, 2473 (1948).

(41) See Reference 4.

(42) See Reference 4.

(43) L. W. Janssen, Rec. Trav. Chim., **65**, 564 (1946).

(44) See Reference 34.

(45) See Reference 15.

(46) See Reference 31.

(47) J. A. Stratton, "Electromagnetic Theory," McGraw- Hill, Inc., New York, 1941, pp 205-207.

(48) J. T. Edward, Chem. Ind., 774 (1956).

(49) S. G. Schultz and A. K. Soloman, J. Gen. Physiol., **44**, 1189 (1961).

Lecture Notes in Physics

Bisher erschienen / Already published

Vol. 1: J. C. Erdmann, Wärmeleitung in Kristallen, theoretische Grundlagen und fortgeschrittene experimentelle Methoden. 1969. DM 20,–

Vol. 2: K. Hepp, Théorie de la renormalisation. 1969. DM 18,–

Vol. 3: A. Martin, Scattering Theory: Unitarity, Analyticity and Crossing. 1969. DM 16,–

Vol. 4: G. Ludwig, Deutung des Begriffs physikalische Theorie und axiomatische Grundlegung der Hilbertraumstruktur der Quantenmechanik durch Hauptsätze des Messens. 1970. DM 28,–

Vol. 5: M. Schaaf, The Reduction of the Product of Two Irreducible Unitary Representations of the Proper Orthochronous Quantummechanical Poincaré Group. 1970. DM 16,–

Vol. 6: Group Representations in Mathematics and Physics. Edited by V. Bargmann. 1970. DM 24,–

Vol. 7: R. Balescu, J. L. Lebowitz, I. Prigogine, P. Résibois, Z. W. Salsburg, Lectures in Statistical Physics. 1971. DM 18,–

Vol. 8: Proceedings of the Second International Conference on Numerical Methods in Fluid Dynamics. Edited by M. Holt. 1971. DM 28,–

Vol. 9: D. W. Robinson, The Thermodynamic Pressure in Quantum Statistical Mechanics. 1971. DM 16,–

Vol. 10: J. M. Stewart, Non-Equilibrium Relativistic Kinetic Theory. 1971. DM 16,–

Vol. 11: O. Steinmann, Perturbation Expansions in Axiomatic Field Theory. 1971. DM 16,–

Vol. 12: Statistical Models and Turbulence. Edited by M. Rosenblatt and C. Van Atta. 1972. DM 28,–

Vol. 13: M. Ryan, Hamiltonian Cosmology. 1972. DM 18,–

Vol. 14: Methods of Local and Global Differential Geometry in General Relativity. Edited by D. Farnsworth, J. Fink, J. Porter and A. Thompson. 1972. DM 18,–

Vol. 15: M. Fierz, Vorlesungen zur Entwicklungsgeschichte der Mechanik. 1972. DM 16,–

Vol. 16: H.-O. Georgii, Phasenübergang 1. Art bei Gittergasmodellen. 1972. DM 18,–

Vol. 17: Strong Interaction Physics. Edited by W. Rühl and A. Vancura. 1973. DM 28,–

Vol. 18: Proceedings of the Third International Conference on Numerical Methods in Fluid Mechanics, Vol. I. Edited by H. Cabannes and R. Temam. 1973. DM 18,–

Vol. 19: Proceedings of the Third International Conference on Numerical Methods in Fluid Mechanics, Vol. II. Edited by H. Cabannes and R. Temam. 1973. DM 26,–

Vol. 20: Statistical Mechanics and Mathematical Problems. Edited by A. Lenard. 1973. DM 22,–

Vol. 21: Optimization and Stability Problems in Continuum Mechanics. Edited by P. K. C. Wang. 1973. DM 16,–

Vol. 22: Proceedings of the Europhysics Study Conference on Intermediate Processes in Nuclear Reactions. Edited by N. Cindro, P. Kulišić and Th. Mayer-Kuckuk. 1973. DM 26,–

Vol. 23: Nuclear Structure Physics. Proceedings of the Minerva Symposium on Physics. Edited by U. Smilansky, I. Talmi, and H. A. Weidenmüller. 1973. DM 26,–

Vol. 24: R. F. Snipes, Statistical Mechanical Theory of the Electrolytic Transport of Non-electrolytes. 1973. DM 20,–